CONTINUED FRACTIONS

by

A. Ya. Khinchin

DOVER PUBLICATIONS, INC.
Mineola, New York

Copyright

Bibliographical Note

This Dover edition, first published in 1997, is an unabridged and unaltered republication of the edition published by The University of Chicago Press in 1964. The first Russian edition was published in 1935. The third Russian Edition, of which this edition is a translation, was published in 1961 by The State Publishing House of Physical-Mathematical Literature, Moscow. This edition was translated from the Russian by Scripta Technica, Inc. The English translation was edited by Herbert Eagle, Brown University.

Library of Congress Cataloging-in-Publication Data

Khinchin, Aleksandr I͡Akovlevich, 1894–1959.
    [T͡Sepnye drobi. English]
    Continued fractions / by A. Ya. Khinchin ; [translated from the Russian by Scripta Technica. Inc.].
       p.   cm.
    Originally published: Chicago : University of Chicago Press, 1964.
    "The English translation was edited by Herbert Eagle"—T.p. verso.
    Includes index.
    ISBN-13: 978-0-486-69630-0
    ISBN-10: 0-486-69630-8 (pbk.)
    1. Continued fractions. I. Eagle, Herbert. II. Title.
QA295.K513   1997
515'.243—dc21
                                                       97-8056
                                                          CIP

Manufactured in the United States by Courier Corporation
69630805
www.doverpublications.com

PREFACE
TO THE THIRD EDITION

The present (third) edition of this outstanding book by A. Ya. Khinchin was undertaken by the State Press for Physics and Mathematics only after the death of its author. For that reason, the book is being reprinted without change except for brief remarks of a bibliographic nature, indicated by my initials (B. G.).

Although it has been more than a quarter of a century since Khinchin wrote this book, it still maintains its original freshness. It is, thus, not without reason that many editions of the book have been printed in various countries in the last decade. Furthermore, in connection with the development of new methods of computational technology, a natural interest has arisen in computational algorithms, including that of continued fractions. Along these lines, there is a useful monograph by A. N. Khovanskii, *Prilozhenie tsepnykh drobei i ikh obobshchenii k voprosam priblizhennogo analiza* ("The Application of Continued Fractions and Their Generalizations to Problems in Approximative Analysis"), published in 1956. Although Khinchin had a different purpose, his book can still serve as an excellent introduction both to the study of the algorithms of continued fractions and to the profound and interesting problems of the measure theory of numbers. Khinchin devoted much time and displayed much initiative in the development of this theory. To a considerable degree, the entire third chapter is the result of his investigations.

I am hoping that this book will be widely read with the same fascination with which many persons, one of them being the author of these lines, read it twenty-five years ago.

B. V. GNEDENKO

PREFACE
TO THE SECOND EDITION

This edition is a reprint of the first with no significant changes. No other monographs on continued fractions have appeared in Russian since publication of the first edition. Among the more general works on the theory of numbers that contain information on continued fractions, the works of D. A. Glava, V. A. Venkov, and I. V. Arnold may be mentioned.

A. KHINCHIN

October 1949

FROM THE PREFACE
TO THE FIRST EDITION

The theory of continued fractions deals with a special algorithm that is one of the most important tools in analysis, probability theory, mechanics, and, especially, number theory. The purpose of the present elementary text is to acquaint the reader only with the so-called *regular* continued fractions, that is, those of the form

$$a_0 + \cfrac{1}{a_1 + \cfrac{1}{a_2 + \dots}}$$

usually with the assumption that all the elements a_i $(i \geq 1)$, are positive integers. This most important and, at the same time, most thoroughly studied class of continued fractions is at the basis of almost all arithmetic and a good many analytic applications of the theory.

I feel that an elementary monograph on the theory of continued fractions is necessary because this theory, which formerly was a part of the mathematical program at the intermediate level, has now been dropped from that program, and hence is no longer included in the new textbooks on elementary algebra. On the other hand, the curricula at the more advanced levels (even in the mathematics divisions of universities) also omit this theory.

Since the basic purpose of this monograph is to fill the gap in our textbook literature, it necessarily had to be elementary and, to as great a degree as possible, accessible. Its *style* is in large measure determined by this fact. Its *content*, however, goes somewhat beyond the limits of that minimum absolutely necessary for any application. This remark applies chiefly to the entire last chapter, which contains the fundamentals of the measure (or probability) theory of continued fractions—an important new field developed almost entirely by Soviet mathematicians; it also applies to quite a number of items in the second chapter, where I attempted, to the extent possible in such an elementary framework, to emphasize the basic role of the apparatus of continued fractions in the study of the arithmetic nature of irrational numbers. I felt that if the fundamentals of the theory of continued fractions were going to be published in the form of a separate monograph, it would be a shame to leave unmentioned those highlights of the theory which are the subject of the greatest amount of contemporary study.

As regards the arrangement of the material, it need only be mentioned that the "formal" part of the study is contained in a special preliminary chapter. In this chapter, the elements of the continued fractions are assumed to be arbitrary positive numbers (not necessarily integers) and often—even more generally—simply independent variables. A drawback to such a separate presentation is the fact that the formal properties of the apparatus being studied are submitted to the reader before the subject matter itself and, therefore, are divorced from it. This is no doubt undesirable from a pedagogical standpoint.

However, a greater methodological precision is to be attained by this approach (because the reader can see immediately which properties of continued fractions come from the very structure of the apparatus and which exist only under the assumption of positive integral elements). Such a separate introductory exposition of the formal part of the study also makes possible the subsequent development of the arithmetic theory (which is the main theme of the study) on an already prepared formal base. Thus, the reader's attention may be concentrated on the content of the material being expounded, without diverting it for purely formal considerations.

A. KHINCHIN

Moscow
February 12, 1935

CONTENTS

Chapter I

PROPERTIES OF THE
APPARATUS

---

1. Introduction

An expression of the form

$$a_0 + \cfrac{1}{a_1 + \cfrac{1}{a_2 + \ldots}} \tag{1}$$

is called a *regular* or *simple* continued fraction. The letters a_0, a_1, a_2, $\cdots$, in the most general treatment of the subject, denote independent variables. In particular cases, these variables may be allowed to take values only in certain specified domains. Thus, a_0, a_1, a_2, $\cdots$ may be assumed to be real or complex numbers, functions of one or several variables, and so on. For the purposes of the present book, we shall always assume a_1, a_2, $\cdots$ to be *positive integers*; a_0 may be an arbitrary real number. We shall call these numbers the *elements* of the given continued fraction. The number of elements may be either finite or infinite. In the first case, we shall write the given continued fraction in the form

$$a_0 + \cfrac{1}{a_1 + \cfrac{1}{a_2 + \ldots \atop {} + \cfrac{1}{a_n}}} \tag{2}$$

and call it a *finite* continued fraction—more precisely an nth-order continued fraction (so that an nth-order continued fraction has $n + 1$ elements); in the second case, we shall write the continued fraction in the form (1) and call it an *infinite* continued fraction.

Every finite continued fraction is the result of a finite number of rational operations on its elements. Therefore, under our assumptions regarding the elements, every finite continued fraction is equal to some real number. In particular, if all the elements are rational numbers, the fraction itself will be a rational number. On the other hand, we cannot immediately assign any numerical values to an infinite continued fraction. Until we adopt some convention, it is only a formal notation, similar to that for an infinite series whose convergence or divergence is not brought into question. Of course, it can, nonetheless, be the subject of mathematical investigations.

Let us agree for reasons of technical convenience to write the infinite continued fraction (1) in the form

$$[a_0; \ a_1, \ a_2, \ \ldots], \tag{3}$$

and the finite continued fraction (2) in the form

$$[a_0; \ a_1, \ a_2, \ \ldots, \ a_n]; \tag{4}$$

thus, the order of a finite continued fraction is equal to the number of symbols (elements) after the semicolon.

Let us agree to call the continued fraction

$$s_k = [a_0; \ a_1, \ a_2, \ \ldots, \ a_k],$$

where $0 \leq k \leq n$, a *segment* of the continued fraction (4). Similarly, for arbitrary $k \geq 0$, we shall call s_k a segment of the infinite continued fraction (3). Obviously, any segment of any continued fraction (finite or infinite) is itself a finite continued fraction. Let us also agree to call the continued fraction

$$r_k = [a_k; \ a_{k+1}, \ \ldots, \ a_n]$$

a *remainder* of the finite continued fraction (4). Similarly, we shall call the continued fraction

$$r_k = [a_k; \ a_{k+1}, \ a_{k+2}, \ \ldots]$$

a remainder of the infinite continued fraction (3). Obviously, all the remainders of a finite continued fraction are finite continued fractions and all the remainders of an infinite continued fraction are infinite continued fractions.

For finite continued fractions, it follows that

$$\begin{aligned} [a_0; \ a_1, \ a_2, \ \ldots, \ a_n] \\ = [a_0; \ a_1, \ a_2, \ \ldots, \ a_{k-1}, \ r_k] \ (0 \leqslant k \leqslant n). \end{aligned} \tag{5}$$

The analogous relationship

$$[a_0; \ a_1, \ a_2, \ \ldots] = [a_0; \ a_1, \ a_2, \ \ldots, \ a_{k-1}, \ r_k] \qquad (k \geqslant 0)$$

for infinite continued fractions can be meaningful only as a formal (trivial) notation since the element r_k on the right side of this equation, being an infinite continued fraction, has no definite numerical value.

2. Convergents

Every finite continued fraction,

$$[a_0; \; a_1, \; a_2, \; \ldots, \; a_n]$$

being the result of a finite number of rational operations on its elements, is a rational function of these elements and, consequently, can be represented as the ratio of two polynomials

$$\frac{P(a_0, \; a_1, \; \ldots, \; a_n)}{Q(a_0, \; a_1, \; \ldots, \; a_n)} \;,$$

in $a_0, a_1, \cdots, a_n$, with integral coefficients. If the elements have numerical values, the given continued fraction is then represented in the form of an ordinary fraction p/q. However, such a representation is, of course, not unique. For what follows, it will be important for us to have a *definite* representation of a finite continued fraction in the form of a simple fraction—a representation which we shall call *canonical*. We shall define such a representation by induction.

For a zeroth-order continued fraction,

$$[a_0] = a_0 \,,$$

we take as our canonical representation the fraction $a_0/1$. Suppose now that canonical representations are defined for continued fractions of order less than n. By equation (5), an nth-order fraction

$$[a_0; \; a_1, \; \ldots, \; a_n] = [a_0; \; r_1] = a_0 + \frac{1}{r_1} \cdot$$

Here,

$$r_1 = [a_1; \; a_2, \; \ldots, \; a_n]$$

is an $(n-1)$st-order continued fraction, for which, consequently, the canonical representation is already defined. Let us represent it as

$$r_1 = \frac{p'}{q'} \;;$$

then,

$$[a_0; \; a_1, \; \ldots, \; a_n] = a_0 + \frac{q'}{p'} = \frac{a_0 p' + q'}{p'} \cdot$$

We shall take this last fraction as our canonical representation of the continued fraction $[a_0; a_1, \cdots, a_n]$. Thus, by setting

$$[a_0; a_1, \ldots, a_n] = \frac{p}{q},$$

$$r_1 = [a_1; a_2, \ldots, a_n] = \frac{p'}{q'},$$

we have the following expressions for the numerators and denominators of these canonical representations:

$$p = a_0 p' + q', \qquad q = p'. \tag{6}$$

Thus, we have uniquely defined canonical representations of continued fractions of all orders.

In the theory of continued fractions, an especially important role is played by the canonical representations of the segments of a given (finite or infinite) continued fraction $a = [a_0; a_1, a_2, \cdots]$. We shall denote by p_k/q_k the canonical representation of the segment

$$s_k = [a_0; a_1, a_2, \ldots, a_k]$$

of the continued fraction, and we shall call it the kth-order *convergent* (or *approximant*) of the continued fraction a. This concept is defined in exactly the same way for finite and infinite continued fractions. The only difference is that a finite continued fraction has a finite number of convergents, whereas an infinite continued fraction has an infinite number of them. For an nth-order continued fraction a, obviously

$$\frac{p_n}{q_n} = a;$$

such a continued fraction has $n + 1$ convergents (of orders, 0, 1, 2, $\cdots$, n).

THEOREM 1 (the rule for the formation of the convergents). *For arbitrary $k \geq 2$,*

$$\left.\begin{array}{l} p_k = a_k p_{k-1} + p_{k-2}, \\ q_k = a_k q_{k-1} + q_{k-2}. \end{array}\right\} \tag{7}$$

PROOF. In the case of $k = 2$, the formulas in (7) are easily verified directly. Let us suppose that they are true for all $k < n$. Let us then consider the continued fraction

$$[a_1; a_2, \ldots, a_n]$$

and let us denote by p'_r/q'_r its rth-order convergent. On the basis of the formulas in (6),

$$p_n = a_0 p'_{n-1} + q'_{n-1},$$
$$q_n = p'_{n-1}.$$

And since, by hypothesis,

$$p'_{n-1} = a_n p'_{n-2} + p'_{n-3},$$
$$q'_{n-1} = a_n q'_{n-2} + q'_{n-3}$$

(here, we have a_n rather than a_{n-1} because the fraction $[a_1; a_2, \cdots, a_n]$ begins with a_1 and not with a_0), it follows on the basis of (6) that

$$p_n = a_0 (a_n p'_{n-2} + p'_{n-3}) + (a_n q'_{n-2} + q_{n-3})$$
$$= a_n (a_0 p'_{n-2} + q'_{n-2}) + (a_0 p'_{n-3} + q_{n-3})$$
$$= a_n p_{n-1} + p_{n-2},$$
$$q_n = a_n p'_{n-2} + p'_{n-3} = a_n q_{n-1} + q_{n-2},$$

which completes the proof.

These recursion formulas (7), which express the numerator and denominator of an nth-order convergent in terms of the element a_n and the numerators and denominators of the two preceding convergents, serve as the formal basis of the entire theory of continued fractions.

REMARK. It is sometimes convenient to consider a convergent of order -1; in this case, we set $p_{-1} = 1$ and $q_{-1} = 0$. Obviously, with this convention (and only then), the formulas of (7) retain their validity for $k = 1$.

THEOREM 2. *For all $k \geq 0$,*

$$q_k p_{k-1} - p_k q_{k-1} = (-1)^k. \tag{8}$$

PROOF. Multiplying the first formula of (7) by q_{k-1} and the second by p_{k-1} and then subtracting the first from the second, we obtain

$$q_k p_{k-1} - p_k q_{k-1} = -(q_{k-1} p_{k-2} - p_{k-1} q_{k-2}),$$

and since

$$q_0 p_{-1} - p_0 q_{-1} = 1,$$

the theorem is proved.

COROLLARY. *For all* $k \geq 1$,

$$\frac{p_{k-1}}{q_{k-1}} - \frac{p_k}{q_k} = \frac{(-1)^k}{q_k q_{k-1}}. \tag{9}$$

THEOREM 3. *For all* $k \geq 1$,

$$q_k p_{k-2} - p_k q_{k-2} = (-1)^{k-1} a_k.$$

PROOF. By multiplying the first formula of (7) by q_{k-2} and the second by p_{k-2} and then subtracting the first from the second, we obtain, on the basis of Theorem 2,

$$q_k p_{k-2} - p_k q_{k-2} = a_k (q_{k-1} p_{k-2} - p_{k-1} q_{k-2}) = (-1)^{k-1} a_k,$$

which completes the proof.

COROLLARY. *For all* $k \geq 2$,

$$\frac{p_{k-2}}{q_{k-2}} - \frac{p_k}{q_k} = \frac{(-1)^{k-1} a_k}{q_k q_{k-2}}. \tag{10}$$

The simple results that we have just obtained make it easy for us to reach certain very important conclusions concerning the relative values of the convergents of a given continued fraction. Specifically, (10) shows that the convergents of even order form an increasing sequence and that those of odd order form a decreasing sequence. Thus, these two sequences tend toward each other (all this under our assumption that the elements from a_1 on are positive). Since, by (9), every odd-order convergent is greater than the immediately following even-order convergent, it follows that every odd-order convergent is greater than any even-order convergent. Therefore, we may draw the following conclusions.

THEOREM 4. *Even-order convergents form an increasing and odd-order convergents a decreasing sequence. Also, every odd-order convergent is greater than any even-order convergent.*

It is particularly evident that, for a finite continued fraction α, every even-order convergent is less than α and every odd-order convergent is greater than α (except, of course, the last convergent, which is equal to α).

We conclude this section with the proof of two simple, but extremely important, propositions concerning the numerators and denominators of the convergents.

THEOREM 5. *For arbitrary k ($1 \leq k \leq n$),*

$$[a_0; a_1, a_2, \ldots, a_n] = \frac{p_{k-1}r_k + p_{k-2}}{q_{k-1}r_k + q_{k-2}}. \qquad (11)$$

(Here, p_i, q_i, r_i refer to the continued fraction on the left side of this equation.)

PROOF. From (5),

$$[a_0; a_1, a_2, \ldots, a_n] = [a_0; a_1, a_2, \ldots, a_{k-1}, r_k].$$

The continued fraction on the right side of this equation has as a $(k-1)$st-order convergent the fraction p_{k-1}/q_{k-1}. Its kth-order convergent, p_k/q_k, is equal to the fraction itself; and since from (7)

$$p_k = p_{k-1}r_k + p_{k-2}, \qquad q_k = q_{k-1}r_k + q_{k-2},$$

the theorem is proved.

THEOREM 6. *For arbitrary $k \geq 1$,*

$$\frac{q_k}{q_{k-1}} = [a_k; a_{k-1}, \ldots, a_1].$$

PROOF. For $k = 1$, this relationship is obvious because it is of the form

$$\frac{q_1}{q_0} = a_1.$$

Suppose that $k > 1$ and that

$$\frac{q_{k-1}}{q_{k-2}} = [a_{k-1}; a_{k-2}, \ldots, a_1]. \qquad (12)$$

On the basis of the equations in (7),

$$q_k = a_k q_{k-1} + q_{k-2}$$

and we have

$$\frac{q_k}{q_{k-1}} = a_k + \frac{q_{k-2}}{q_{k-1}} = \left[a_k; \frac{q_{k-1}}{q_{k-2}}\right].$$

Therefore, from formulas (5) and (12),

$$\frac{q_k}{q_{k-1}} = [a_k; a_{k-1}, \ldots, a_1],$$

which completes the proof.

3. *Infinite continued fractions*

To every infinite continued fraction

$$[a_0; a_1, a_2, \ldots],\tag{13}$$

there corresponds an infinite sequence of convergents

$$\frac{p_0}{q_0}, \frac{p_1}{q_1}, \ldots, \frac{p_k}{q_k}, \ldots. \tag{14}$$

Every convergent is some real number. If the sequence (14) converges, that is, if it has a unique limit a, it is natural to consider this number a as the "value" of the continued fraction (13) and to write

$$a = [a_0; a_1, a_2, \ldots].$$

The continued fraction (13) itself is then said to *converge*. If the sequence (14) does not have a definite limit, we say that the continued fraction (13) diverges.

In many of their properties, convergent infinite continued fractions are analogous to finite continued fractions. The basic property which makes possible the further extension of this analogy is expressed by the following theorem.

THEOREM 7. *If the infinite continued fraction* (13) *converges, so do all of its remainders; conversely, if at least one of the remainders of the continued fraction* (13) *converges, the continued fraction itself converges.*

PROOF. Let us agree to denote by p_k/q_k the convergents of a given continued fraction (13), and by p'_k/q'_k the convergents of any one of its remainders, for example, r_n. From formula (11), we have

$$\frac{p_{n+k}}{q_{n+k}} = [a_0; a_1, a_2, \ldots, a_{n+k}] = \frac{p_{n-1}\dfrac{p'_k}{q'_k} + p_{n-2}}{q_{n-1}\dfrac{p'_k}{q'_k} + q_{n-2}} \quad (k = 0, 1, \ldots). \tag{15}$$

It follows immediately that if the remainder r_n converges, that is, if as $k \to \infty$ the fraction p'_k/q'_k approaches a limit which we shall also denote by r_n, then the fraction p_{n+k}/q_{n+k} will converge to a limit a equal to

$$a = \frac{p_{n-1}r_n + p_{n-2}}{q_{n-1}r_n + q_{n-2}}. \tag{16}$$

By solving (15) for p'_k/q'_k, we establish the validity of the converse, thus completing the proof of the theorem.

We note that formula (16), which we have just established for convergent infinite continued fractions, is exactly analogous to formula (11), which we proved earlier for finite continued fractions. Similarly, the theorem analogous to Theorem 5 holds for infinite continued fractions.

The following propositions for convergent infinite continued fractions follow directly from Theorem 4 of the preceding section.

THEOREM 8. *The value of a convergent infinite continued fraction is greater than any of its even-order convergents and is less than any of its odd-order convergents.*

Furthermore, on the basis of this theorem, the corollary to Theorem 2 of the preceding section implies the following result, which plays a basic role in the arithmetic applications of the theory of continued fractions.

THEOREM 9. *The value α of the convergent infinite continued fraction (13) for arbitrary $k \geq 0$ satisfies the inequality*[1]

$$\left| \alpha - \frac{p_k}{q_k} \right| < \frac{1}{q_k q_{k+1}} .$$

Obviously, Theorem 9 is also valid for the finite continued fraction

$$\alpha = [a_0; a_1, a_2, \ldots, a_n] ,$$

for all $k < n$, except that, for the single case of $k = n - 1$, the inequality must be replaced by equality, since $\alpha = p_n/q_n$. If α is the value of a convergent infinite continued fraction (13), we shall also refer to the elements of that continued fraction as the *elements of the number* α. Similarly, we shall refer to the convergents, segments, and remainders of the continued fraction (13) as the convergents, segments, and remainders, respectively, of the number α. On the basis of Theorem 7, all the remainders of a convergent infinite continued fraction (13) have definite real values.

The question naturally arises as to whether there are tests for the convergence of continued fractions, just as for infinite series. In the case with which we are concerned, that is, when $a_i > 0$, for all $i \geq 1$, there exists an extremely simple and convenient test for convergence.

---

[1] We note that, under our assumptions, $q_k > 0$, for all $k \geq 0$ (since $q_0 = 1$ and $q_1 = a_1$, we can show by induction from the second part of eq. [7] that $q_k > 0$, for all $k > 1$).

THEOREM 10. *For the continued fraction (13) to converge, it is necessary and sufficient that the series*

$$\sum_{n=1}^{\infty} a_n \qquad (17)$$

diverge.

PROOF. It clearly follows from Theorem 4 that a necessary and sufficient condition for the convergence of an infinite continued fraction is that the two sequences referred to in that theorem have the same limit. (Theorem 4 clearly implies that each of these sequences has a limit.) And, as formula (9) shows, this is the case if and only if

$$q_k q_{k+1} \to \infty \quad \text{as} \quad k \to \infty. \qquad (18)$$

Thus, condition (18) is necessary and sufficient for the convergence of a given continued fraction.

Suppose that the series (17) converges. From the second formula of (7),

$$q_k > q_{k-2} \qquad (k \geqslant 1).$$

Therefore, for arbitrary k, we have either $q_k > q_{k-1}$ or $q_{k-1} > q_{k-2}$. In the first case, the second formula of (7) yields

$$q_k < a_k q_k + q_{k-2},$$

and therefore, for sufficiently large k (when $a_k < 1$, which, because of the convergence of the series in eq. [17], must be the case for $k \geq k_0$), we have

$$q_k < \frac{q_{k-2}}{1-a_k}.$$

In the second case, the same formula gives, for $a_k < 1$,

$$q_k < (1+a_k) q_{k-1} < \frac{q_{k-1}}{1-a_k}.$$

Thus, for all $k \geq k_0$, we have

$$q_k < \frac{1}{1-a_k} q_l,$$

where $l < k$. If $l \geq k_0$, we may apply the same inequality to q_l.

By continuing this reasoning, we arrive at the inequality

$$q_k < \frac{q_s}{(1-a_k)(1-a_l)\ldots(1-a_r)}, \tag{19}$$

where $k > l > \cdots > r \geq k_0$ and $s < k_0$. But, because of the assumed convergence of the series in (17), the infinite product

$$\prod_{n=k_0}^{\infty}(1-a_n),$$

as we know, converges: that is, it has a positive value, which we denote by λ. Obviously,

$$(1-a_k)(1-a_l)\ldots(1-a_r) \geqslant \prod_{n=k_0}^{k}(1-a_n)=\lambda.$$

Therefore, if we denote by Q the largest of the numbers $q_0, q_1, \cdots,$ q_{k_0-1}, we conclude from inequality (19) that

$$q_k < \frac{Q}{\lambda} \qquad (k \geqslant k_0),$$

consequently,

$$q_{k+1}q_k < \frac{Q^2}{\lambda^2} \qquad (k \geqslant k_0),$$

and the relationship in (18) cannot hold. Therefore, the given continued fraction diverges.

Conversely, suppose that the series in (17) diverges. Since $q_k > q_{k-2}$, for all $k \geq 2$, if we denote by c the smallest of the numbers q_0, q_1, we have $k \geq 0$, for arbitrary $q_k \geq c$. Therefore, the second formula of (7) gives us

$$q_k \geqslant q_{k-2}+ca_k \qquad (k \geqslant 2).$$

Successive application of this inequality gives us

$$q_{2k} \geqslant q_0 + c \sum_{n=1}^{k} a_{2n}$$

and

$$q_{2k+1} \geqslant q_1 + c \sum_{n=1}^{k} a_{2n+1},$$

so that

$$q_{2k} + q_{2k+1} > q_0 + q_1 + c \sum_{n=1}^{2k+1} a_n;$$

in other words, for all k,

$$q_k + q_{k-1} > c \sum_{n=1}^{k} a_n.$$

We have already proved this inequality for odd values of k, and it can obviously be established for even values of k by the same method. It then follows that at least one of the factors in the product $q_k q_{k-1}$ exceeds $\frac{1}{2} c \sum_{n=1}^{k} a_n$, and since the other factor is never less than c, we have

$$q_k q_{k-1} > \frac{c^2}{2} \sum_{n=1}^{k} a_n.$$

Because of the assumed divergence of the series in (17), this implies relationship (18) and, consequently, the convergence of the given continued fraction. This completes the proof of Theorem 10.

4. Continued fractions with natural elements

From this point until the end of the book, we shall assume that elements a_1, a_2, $\cdots$ are natural numbers, that is, positive integers, and that a_0 is an integer, though not necessarily positive. If such a continued fraction is infinite, Theorem 10 ensures its convergence. Therefore, we can henceforth freely assume that any continued fraction that we are dealing with is convergent, and we can speak of its "value". If such a continued fraction is finite, and if its last element (a_n) is 1, it is evident that $r_{n-1} = a_{n-1} + 1$ is an integer. Therefore, in this case, we can write the given nth-order continued fraction $[a_0; a_1, a_2, \cdots, a_{n-1}, 1]$ in the form of an $(n-1)$st-order continued fraction $[a_0; a_1, a_2, \cdots, a_{n-1} + 1]$; in this new form, the last element is clearly greater than unity.

Because of this fact, in all that follows we can exclude from consideration finite continued fractions whose last elements are equal to unity (except, of course, for the zeroth-order fraction [1]). This plays an important role in the question of the uniqueness of the representation of numbers by continued fractions (see Chap. II, sec. 5).

Obviously, the numerators and denominators of the convergents, in the case now under consideration, are integers. (For p_{-1}, q_{-1}, p_0, and q_0, this can be seen immediately, and for the numerators and denominators of the remaining convergents, it follows from the formulas in eq. [7].) Furthermore, we have the following very important proposition.

THEOREM 11. *All convergents are irreducible.*

The proof follows immediately from formula (8), since any common divisor of the numbers p_n and q_n would at the same time be a divisor of the expression $q_n p_{n-1} - p_n q_{n-1}$.

The second formula of (7) shows that $q_k > q_{k-1}$, for every $k \geq 2$. Therefore, the sequence

$$q_1, q_2, \ldots, q_k, \ldots$$

is always increasing. We have a much stronger proposition concerning the rate of increase of the numbers q_k.

THEOREM 12. *For arbitrary*[2] *$k \geq 2$,*

$$q_k \geq 2^{\frac{k-1}{2}}.$$

PROOF. For $k \geq 2$,

$$q_k = a_k q_{k-1} + q_{k-2} \geq q_{k-1} + q_{k-2} \geq 2q_{k-2}.$$

Successive application of this inequality yields

$$q_{2k} \geq 2^k q_0 = 2^k, \qquad q_{2k+1} \geq 2^k q_1 \geq 2^k,$$

which proves the theorem. Thus, the denominators of the convergents increase at least as rapidly as the terms of a geometric progression.

Intermediate fractions.—Suppose that $k \geq 2$ and that i is an arbitrary negative integer. The difference

$$\frac{p_{k-1}(i+1) + p_{k-2}}{q_{k-1}(i+1) + q_{k-2}} - \frac{p_{k-1}i + p_{k-2}}{q_{k-1}i + q_{k-2}},$$

which, as is easily seen, is equal to

$$\frac{(-1)^k}{[q_{k-1}(i+1) + q_{k-2}][q_{k-1}i + q_{k-2}]},$$

has the same sign for all $i \geq 0$, depending only on whether k is even or odd. It follows from this that the fractions

$$\frac{p_{k-2}}{q_{k-2}}, \frac{p_{k-2} + p_{k-1}}{q_{k-2} + q_{k-1}}, \frac{p_{k-2} + 2p_{k-1}}{q_{k-2} + 2q_{k-1}}, \ldots, \frac{p_{k-2} + a_k p_{k-1}}{q_{k-2} + a_k q_{k-1}} = \frac{p_k}{q_k} \quad (20)$$

[2] Here, and in all that follows, in the case of a *finite* continued fraction only those values of k for which q_k is meaningful are to be considered.

form, for even k, an increasing and, for odd k, a decreasing sequence (see Theor. 4). The first and last terms of this sequence are either both even- or both odd-order convergents. The intervening terms (if there are any, that is, if $a_k > 1$), we shall call *intermediate fractions*. In arithmetic applications, these intermediate fractions play an important role (though not as important a role as the convergents). To make their mutual disposition and the law of their progressive formation clearer, it is convenient to introduce the concept of the so-called *mediant* of two fractions.

The mediant of two fractions a/b and c/d, with positive denominators, is the fraction

$$\frac{a+c}{b+d}.$$

LEMMA. *The mediant of two fractions always lies between them in value.*

PROOF. Suppose, for definiteness, that $a/b \le c/d$. Then, $bc - ad \ge 0$, and, consequently,

$$\frac{a+c}{b+d} - \frac{a}{b} = \frac{bc - ad}{b(b+d)} \ge 0, \qquad \frac{a+c}{b+d} - \frac{c}{d} = \frac{ad - bc}{b(b+d)} \le 0,$$

which proves the lemma.

We see immediately that each of the intermediate fractions in the progression of (20) is the mediant of the preceding fraction and the fraction p_{k-1}/q_{k-1}. By going through progression (20) and successively forming the mediants, we proceed from the convergents p_{k-2}/q_{k-2} in the direction of the convergents p_{k-1}/q_{k-1}. The concluding step in this sequence will occur when the mediant constructed coincides with p_k/q_k. This last fraction lies between p_{k-1}/q_{k-1} and p_{k-2}/q_{k-2}, as we know from Theorem 4. We also know that the value a of the given continued fraction lies between p_{k-1}/q_{k-1} and p_k/q_k, and that the fractions p_{k-2}/q_{k-2} and p_k/q_k, which are either both of even order or both of odd order, lie on the same side of the number a. It follows from this that the entire progression in (20) lies on one side of the number a and that the fraction p_{k-1}/q_{k-1} lies on the other side. In particular, the fractions $(p_{k-1} + p_{k-2})/(q_{k-1} + q_{k-2})$ and p_{k-1}/q_{k-1} are always on opposite sides of a. In other words, *the value of a continued fraction always lies between an arbitrary convergent and the mediant of that convergent and the preceding one.* (We suggest that the reader make a drawing to illustrate the relative positions of all these numbers.)

This remark indicates a method whereby, if we know the convergents p_{k-2}/q_{k-2} and p_{k-1}/q_{k-1}, we can construct the subsequent convergent p_k/q_k without knowing the element a_k (but using our knowledge of the value α of the continued fraction). Specifically, we first take the mediant of the two given fractions, then the mediant of this mediant and p_{k-1}/q_{k-1}, and so on, each time taking the mediant of the mediant just obtained and the fraction p_{k-1}/q_{k-1}. We already know that these consecutive mediants will initially approximate α. *The last mediant of this progression that lies on the same side of α as does the initial fractions p_{k-2}/q_{k-2} is p_k/q_k.* For, as we already know, p_k/q_k lies somewhere among the mediants in the progression, and on the same side of α as p_{k-2}/q_{k-2}. Therefore, it only remains for us to show that the subsequent mediant will lie on the opposite side of α. But the last mediant is $(p_k + p_{k-1})/(q_k + q_{k-1})$ and, on the basis of the remark made above, it does indeed lie on the opposite side of the number α.

There is another even more important consequence of the relative positions of a number α, and its convergents and intermediate fractions. The intermediate fraction $(p_k + p_{k+1})/(q_k + q_{k+1})$, since it is between p_k/q_k and α, lies closer to p_k/q_k than does α; that is,

$$\left| \alpha - \frac{p_k}{q_k} \right| > \left| \frac{p_k + p_{k+1}}{q_k + q_{k+1}} - \frac{p_k}{q_k} \right| = \frac{1}{q_k(q_k + q_{k+1})}.$$

(Equality is impossible here because this would indicate that

$$\alpha = \frac{p_k + p_{k+1}}{q_k + q_{k+1}} = \frac{p_{k+2}}{q_{k+2}}, \qquad a_{k+2} = 1;$$

that is, that α would be a finite continued fraction with last element equal to unity, which we excluded from consideration in the beginning.)

Thus, we arrive at the following important result.

THEOREM 13. *For all $k \geq 0$,*

$$\left| \alpha - \frac{p_k}{q_k} \right| > \frac{1}{q_k(q_{k+1} + q_k)}. \tag{21}$$

This inequality, which gives a *lower* bound for the difference $|\alpha - (p_k/q_k)|$, supplements the inequality exhibited in Theorem 9, which provides an upper bound for the same difference.

Chapter II

THE REPRESENTATION OF NUMBERS BY CONTINUED FRACTIONS

5. Continued fractions as an apparatus for representing real numbers

THEOREM 14. *To every real number α, there corresponds a unique continued fraction with value equal to α. This fraction is finite if α is rational and infinite if α is irrational.*[1]

PROOF. We denote by a_0 the largest integer not exceeding α. If α is not an integer, the relation

$$\alpha = a_0 + \frac{1}{r_1} \tag{22}$$

allows us to determine the number r_1. Here, clearly, $r_1 > 1$, since

$$\frac{1}{r_1} = \alpha - a_0 < 1.$$

In general, if r_n is not an integer, we denote by a_n the largest integer not exceeding r_n and define the number r_{n+1} by the relation

$$r_n = a_n + \frac{1}{r_{n+1}}. \tag{23}$$

This procedure can be continued as long as r_n is not an integer; here, clearly, $r_n > 1$ ($n \geq 1$).

Equation (22) shows that

$$\alpha = [a_0; r_1].$$

Suppose that, in general,

$$\alpha = [a_0; a_1, a_2, \ldots, a_{n-1}, r_n]. \tag{24}$$

---

[1] We remind the reader that we are considering continued fractions with integral elements, that $a_i > 0$ for $i \geq 1$, and that the last element of every finite continued fraction must be different from unity.

Then, from equations (5) and (23), we have

$$\alpha = [a_0;\ a_1,\ a_2,\ \ldots,\ a_{n-1},\ a_n,\ r_{n+1}];$$

thus, (24) is valid for all n (assuming, of course, that $r_1, r_2, \cdots, r_{n-1}$ are not integers).

If the number α is rational, all the r_n will clearly be rational. It is easy to see that, in this case, our process will stop after a finite number of steps. If, for example, $r_n = a/b$, then

$$r_n - a_n = \frac{a - ba_n}{b} = \frac{c}{b},$$

where $c < b$, since $r_n - a_n < 1$. Equation (23) then gives

$$r_{n+1} = \frac{b}{c}$$

(provided c is not equal to zero, that is, if r_n is not an integer; if r_n is an integer, our assertion is already satisfied). Thus, r_{n+1} has a smaller denominator than does r_n. It follows from this that if we consider $r_1, r_2,$ $\cdots$, we must eventually come to an integer $r_n = a_n$. But, in this case, (24) asserts that the number α is represented by a finite continued fraction, the last element of which is $a_n = r_n > 1$.

If α is irrational, then all the r_n are irrational and our process is infinite. Setting

$$[a_0;\ a_1,\ a_2,\ \ldots,\ a_n] = \frac{p_n}{q_n}$$

(where the fraction p_n/q_n is irreducible and $q_n > 0$), we have, by (24) and (16) of Chapter I,

$$\alpha = \frac{p_{n-1}r_n + p_{n-2}}{q_{n-1}r_n + q_{n-2}} \qquad (n \geqslant 2).$$

On the other hand, it is obvious that

$$\frac{p_n}{q_n} = \frac{p_{n-1}a_n + p_{n-2}}{q_{n-1}a_n + q_{n-2}},$$

so that

$$\alpha - \frac{p_n}{q_n} = \frac{(p_{n-1}q_{n-2} - q_{n-1}p_{n-2})(r_n - a_n)}{(q_{n-1}r_n + q_{n-2})(q_{n-1}a_n + q_{n-2})}$$

and, consequently,

$$\left| \alpha - \frac{p_n}{q_n} \right| < \frac{1}{(q_{n-1}r_n + q_{n-2})(q_{n-1}a_n + q_{n-2})} < \frac{1}{q_n^2}.$$

Thus,

$$\frac{p_n}{q_n} \to \alpha \quad \text{as} \quad n \to \infty;$$

but this means that the infinite continued fraction $[a_0; a_1, a_2, \cdots]$ has as its value the given number α.

Thus, we have shown that any number α can be represented by a continued fraction; this fraction is finite if α is rational and infinite if α is irrational. It remains for us to show the uniqueness of the expansions that we have obtained. We note first that uniqueness follows essentially from the considerations of section 4, Chapter I, where we saw that once we know the value of a given continued fraction we can effectively construct all its convergents and hence all its elements. However, the required uniqueness can be established in a much simpler manner. Suppose that

$$\alpha = [a_0; a_1, a_2, \ldots] = [a_0'; a_1', a_2', \ldots],$$

where the two continued fractions may be either finite or infinite. Let us denote by $[x]$ the largest integer not exceeding x. First of all, it is obvious that $a_0 = [\alpha]$ and $a_0' = [\alpha]$, so that $a_0 = a_0'$. Furthermore, if it is established that

$$a_i = a_i' \qquad (i = 0, 1, 2, \ldots, n),$$

then, in analogous notation,

$$\left.\begin{aligned} p_i &= p_i' \\ q_i &= q_i' \end{aligned}\right\} \qquad (i = 0, 1, 2, \ldots, n),$$

and, on the basis of formula (16) of Chapter I,

$$\alpha = \frac{p_n r_{n+1} + p_{n-1}}{q_n r_{n+1} + q_{n-1}} = \frac{p_n' r_{n+1}' + p_{n-1}'}{q_n' r_{n+1}' + q_{n-1}'} = \frac{p_n r_{n+1}' + p_{n-1}}{q_n r_{n+1}' + q_{n-1}},$$

so that, $r_{n+1} = r_{n+1}'$. Since $a_{n+1} = [r_{n+1}]$ and $a_{n+1}' = [r_{n+1}']$, we have $a_{n+1} = a_{n+1}'$: that is, the two fractions coincide completely.

We note that the above argument would be impossible if we admitted finite continued fractions with the last element equal to unity; if, for example, $a_{n+1} = 1$ were such a last element, we would have $r_n = a_n + 1$ and $a_n \neq [r_n]$.

We have just shown that real numbers are uniquely represented by continued fractions. The basic significance of such a representation consists, of course, in the fact that, knowing the continued fraction that represents a real number, we can determine the value of that number with an arbitrary prestated degree of accuracy. Therefore, the apparatus of continued fractions can, at least in principle, claim a role in the representation of real numbers similar to that, for example, of decimal or of systematic fractions (that is, fractions constructed according to some system of calculation).

What are the basic advantages and shortcomings of continued fractions as a means of representing the real numbers in comparison with the much more widely used systematic representation? To answer this question, we need first to have a clear picture of the demands that may and should be made of such a representation. Clearly, the first and basic *theoretical* demand should be that the apparatus reflect as much as possible the properties of the number that it represents, so that these properties may be brought out as completely and as simply as possible each time that the representation of the number by this apparatus is given.

With respect to this first demand, continued fractions have an undeniable and considerable advantage over systematic (and, in particular, decimal) fractions. We shall gradually see this during the course of the present chapter. To a degree, in fact, this is clear even from a priori considerations. Since a systematic fraction is connected with a certain system of calculation, it therefore unavoidably reflects, not so much the absolute properties of the number that it represents, as its relationship to that particular system of calculation. Continued fractions, on the other hand, are not connected with any system of calculation; they reproduce in a pure form the properties of the number that they represent. Thus, we have already seen that the rationality or irrationality of the number represented finds complete expression in the finiteness or infiniteness of the continued fraction corresponding to it. As we know, for systematic fractions the corresponding test is considerably more complicated: the finiteness or infiniteness of the representing fraction depends not just on the number represented but also, in a very real way, on its relationship to the chosen system of calculation.

However, besides the basic theoretical demands that we have mentioned, certain demands of a *practical* nature should naturally be made for any apparatus that is used to represent numbers. (Some of these practical considerations may also have certain theoretical value.) Thus, it is of great importance that the apparatus make it possible and reasonably easy to find values that approximate the represented number with any arbitrary degree of accuracy. The apparatus of continued fractions satisfies this demand to a very high degree (and, in any case, better than does the apparatus of systematic fractions). In fact, we shall soon see that the approximating values given by continued fractions have, in a certain extremely simple and important sense, the property of being the *best approximations.*

There is, however, another and yet more significant practical demand that the apparatus of continued fractions does not satisfy at all. Knowing the representations of several numbers, we would like to be able, with relative ease, to find the representations of the simpler functions of these numbers (especially, their sum and product). In brief, for an apparatus to be suitable from a practical standpoint, it must admit sufficiently simple rules for arithmetical operations; otherwise, it cannot serve as a tool for calculation. We know how convenient systematic fractions are in this respect. On the other hand, for continued fractions there are no practically applicable rules for arithmetical operations; even the problem of finding the continued fraction for a sum from the continued fraction representing the addends is exceedingly complicated, and unworkable in computational practice.

The advantages and shortcomings of continued fractions as compared with systematic fractions determine (to a great extent) the areas of application of these two representations. Whereas, in computation, systematic fractions are used almost exclusively, the apparatus of continued fractions finds its primary application in theoretical investigations involving the study of the arithmetic laws of the continuum and the arithmetic properties of individual irrational numbers. The apparatus of continued fractions is an irreplaceable tool for theoretical investigations, and the prime purpose of all that follows will be its application to that purpose.

6. *Convergents as best approximations*

To represent an irrational number α as an ordinary rational fraction (to within a specified margin of accuracy), it is natural to use the convergents of the continued fraction representing α. The degree of accuracy of this approximation is given by Theorems 9 and 13 of Chapter I. Specifically, we have

$$\frac{1}{q_n(q_n + q_{n+1})} < \left| \alpha - \frac{p_n}{q_n} \right| \leqslant \frac{1}{q_n q_{n+1}}.$$

The problem of approximating irrational numbers by rational fractions consists, in its simplest form, of determining which of the fractions that differ from the given irrational number by not more than a specified amount has the lowest (positive) denominator. The problem (stated in this manner) can be meaningful even in the case in which the number a is rational. For example, if a is a fraction with an extremely large numerator and denominator, we may want to approximate this number by a fraction whose numerator and denominator are smaller. From a purely practical point of view, there is no real difference between these two cases (rational and irrational a), since, in practice, every number is given with only a certain degree of accuracy.

It is immediately clear that the apparatus of systematic fractions is completely unsuitable for solving this problem, since the denominators of the approximating fraction that it provides are determined exclusively by the chosen system of calculation (in the case of decimal fractions, they are powers of ten); hence, the denominators are completely independent of the arithmetic nature of the number represented. On the other hand, in the case of a continued fraction, the denominators of the convergents are completely determined by the number repre-
of the convergents are completely determined by the number represented. We, therefore, have every reason to expect that these convergents (since they are connected in a close and natural way with the number represented, and are completely determined by it) will play a significant role in the solution of the problem of the best approximation of a number by a rational fraction.[2]

Let us agree to call a rational fraction a/b (for $b > 0$) a *best approximation* of a real number a if every other rational fraction with the same or smaller denominator differs from a by a greater amount, in other words, if the inequalities $0 < d \leq b$, and $a/b \neq c/d$ imply that

$$\left| a - \frac{c}{d} \right| > \left| a - \frac{a}{b} \right|$$

---

[2] Two interesting algorithms for representing irrational numbers were advanced by M. V. Ostrogradskiĭ shortly before his death. His brief notes on the matter were discussed on bits of paper in the manuscript depository of the Academy of Sciences of the Ukrainian SSR. These notes were deciphered in an article by E. Ya. Remez, "O znakoperemennykh ryadakh, kotorye mogut byt'svyazany s dvumya algorifmami M. V. Ostrogradskogo dlya priblizheniya irratsional'nykh chisel" ("Alternating series that may be connected with two algorithms of M. V. Ostrogradskiĭ for approximating irrational numbers"), *Uspekhi matematicheskikh nauk*, 6, No. 5 (45), 33–42 (1951). As Remez discovered, Ostrogradskiĭ's algorithms give better approximations than continued fractions in certain cases. Unfortunately, no detailed study of these algorithms, even for computational purposes, has as yet been made. (B. G.)

THEOREM 15. *Every best approximation of a number* a *is a convergent or an intermediate fraction of the continued fraction representing that number.*

PRELIMINARY REMARK. For this proposition to have no exceptions it is necessary, as we agreed in section 2, to introduce into our considerations convergents of order -1, by setting $p_{-1} = 1$ and $q_{-1} = 0$. For example, the fraction $\frac{1}{3}$ is, as we can easily verify, a best approximation of the number $\frac{1}{4}$; however, it is not one of the convergents or intermediate fractions of that number, since the set of these fractions (if we begin with the convergents of order zero) consists of only two numbers, namely, $\frac{0}{1}$ and $\frac{1}{4}$. However, if we take the fraction $\frac{1}{0}$ as a convergent of order -1, this set will consist of

$$\frac{1}{0}, \ \frac{0}{1}, \ \frac{1}{1}, \ \frac{1}{2}, \ \frac{1}{3}, \ \frac{1}{4},$$

thus including the fraction $\frac{1}{3}$.

PROOF. Suppose that a/b is a best approximation of the number a. Then, first of all, $a/b \geq a_0$, because if $a/b < a_0$, the fraction $a_0/1$ (being distinct from a/b and having a denominator that is no greater than b) would lie closer to a than does a/b. Therefore, a/b would not be a best approximation.

In a similar manner, we can show that

$$\frac{a}{b} \leqslant a_0 + 1.$$

Thus, we know that $a_0 < (a/b) < a_0 + 1$. If $a/b = a_0$ or $a/b = a_0 + 1$, the conclusion of the theorem would be evident since $a_0/1 = p_0/q_0$ is a convergent and $(a_0 + 1)/1 = (p_0 + p_{-1})/(q_0 + q_{-1})$ is an intermediate fraction of a.

If the fraction a/b does not coincide with any convergent or intermediate fraction of the number a, it must lie between two consecutive such fractions. For instance, for properly chosen k and r (with $k > 0$, $0 \leq r < a_{k+1}$ or $k = 0$, $1 \leq r < a_1$), it will lie between the fractions

$$\frac{p_k r + p_{k-1}}{q_k r + q_{k-1}}$$

and

$$\frac{p_k (r+1) + p_{k-1}}{q_k (r+1) + q_{k-1}},$$

so that

$$\left| \frac{a}{b} - \frac{p_k r + p_{k-1}}{q_k r + q_{k-1}} \right| < \left| \frac{p_k(r+1) + p_{k-1}}{q_k(r+1) + q_{k-1}} - \frac{p_k r + p_{k-1}}{q_k r + q_{k-1}} \right|$$

$$= \frac{1}{\{q_k(r+1) + q_{k-1}\}\{q_k r + q_{k-1}\}} \cdot$$

But, on the other hand, it is obvious that

$$\left| \frac{a}{b} - \frac{p_k r + p_{k-1}}{q_k r + q_{k-1}} \right| = \frac{m}{b(q_k r + q_{k-1})},$$

where m is a positive integer and hence is at least equal to unity. Consequently,

$$\frac{1}{b(q_k r + q_{k-1})} < \frac{1}{\{q_k(r+1) + q_{k-1}\}\{q_k r + q_{k-1}\}},$$

and hence,

$$q_k(r+1) + q_{k-1} < b.$$

The fraction

$$\frac{p_k(r+1) + p_{k-1}}{q_k(r+1) + q_{k-1}}, \tag{25}$$

with denominator less than b, is closer to the number a than is the fraction

$$\frac{p_k r + p_{k-1}}{q_k r + q_{k-1}} \tag{26}$$

(because, in general, from the result of sec. 4, every intermediate fraction is closer to a than is the preceding one) and hence, is also closer than is the fraction a/b, which lies between expressions (25) and (26). However, this contradicts the definition of a best approximation, thus proving Theorem 15.

In the definition of the concept of best approximation, which is at the basis of this theorem, we evaluated the closeness of the rational fraction a/b to the number a in terms of the smallness (in absolute value) of the difference $a - (a/b)$ (which, of course, is the most natural procedure). However, it is often more important or convenient in number theory to examine the difference $ba - a$, which differs from the preceding one only by the factor b. Thus, the smallness of this difference (in absolute value) can also serve as a measure of the closeness of the fraction a/b to the number a. This change from one characteristic to another may at first glance seem trivial, and frequently it is.

However, this is not always the case, as we shall soon see. The significant point is that the factor b is not a constant, but is dependent on the approximating fraction itself, and changes when this fraction is changed.

Let us now agree to refer to those best approximations mentioned in Theorem 15 as *best approximations of the first kind*. Let us further agree to call the rational fraction a/b (where $b > 0$) a *best approximation of the second kind* of a number α if the inequalities $c/d \neq a/b$ and $0 < d \le b$ imply

$$| d\alpha - c | > | b\alpha - a |.$$

Best approximations of the second kind are thus defined in terms of the characteristic $|b\alpha - a|$ in a manner completely analogous to the definition of best approximations of the first kind in terms of the characteristic $|\alpha - a/b|$.

It is easy to show that every best approximation of the second kind must necessarily be a best approximation of the first kind. For if

$$\left| \alpha - \frac{c}{d} \right| \le \left| \alpha - \frac{a}{b} \right| \qquad \left(\frac{c}{d} \neq \frac{a}{b}, \ d \le b \right),$$

on multiplying the first of these inequalities by the third, we would obtain

$$| d\alpha - c | \le | b\alpha - a |;$$

in other words, if the fraction a/b was not a best approximation of the first kind, it could not be a best approximation of the second kind.

The converse is not true: a best approximation of the first kind can fail to be a best approximation of the second kind. For example, the fraction $\frac{1}{3}$ can easily be shown to be a best approximation of the first kind of the number $\frac{1}{5}$. However, that it is not a best approximation of the second kind is seen from the inequality

$$\left| 1 \cdot \frac{1}{5} - 0 \right| < \left| 3 \cdot \frac{1}{5} - 1 \right| \qquad (1 < 3).$$

It follows from these remarks and from Theorem 15 that all best approximations of the second kind are convergents or intermediate fractions. However—and here lies the fundamental significance of the apparatus of continued fractions in finding best approximations of the second kind—we can make a much stronger assertion.

THEOREM 16. *Every best approximation of the second kind is a convergent.*

Proof. Suppose that a fraction a/b is a best approximation of the second kind of the number

$$\alpha = [a_0;\ a_1,\ a_2,\ \ldots],$$

whose convergents will be denoted by p_k/q_k. If $a/b < a_0$, we would obtain

$$|1 \cdot \alpha - a_0| < \left|\alpha - \frac{a}{b}\right| \leqslant |b\alpha - a| \qquad (1 \leqslant b),$$

that is, a/b would not be a best approximation of the second kind. Thus, $a/b \geq a_0$. But then the fraction a/b, if it did not coincide with one of the convergents, would either lie between two convergents p_{k-1}/q_{k-1} and p_{k+1}/q_{k+1}, or would be greater than p_1/q_1. In the first case,

$$\left|\frac{a}{b} - \frac{p_{k-1}}{q_{k-1}}\right| \geqslant \frac{1}{bq_{k-1}}$$

and

$$\left|\frac{a}{b} - \frac{p_{k-1}}{q_{k-1}}\right| < \left|\frac{p_k}{q_k} - \frac{p_{k-1}}{q_{k-1}}\right| = \frac{1}{q_k q_{k-1}},$$

so that

$$b > q_k; \qquad (27)$$

on the other hand,

$$\left|\alpha - \frac{a}{b}\right| \geqslant \left|\frac{p_{k+1}}{q_{k+1}} - \frac{a}{b}\right| \geqslant \frac{1}{bq_{k+1}},$$

and hence,

$$|b\alpha - a| \geqslant \frac{1}{q_{k+1}},$$

whereas

$$|q_k\alpha - p_k| \leqslant \frac{1}{q_{k+1}},$$

so that

$$|q_k\alpha - p_k| \leqslant |b\alpha - a|. \qquad (28)$$

Inequalities (27) and (28) show that a/b is not a best approximation of the second kind.

In the second case (that is, if $a/b > p_1/q_1$), we have

$$\left|\alpha - \frac{a}{b}\right| > \left|\frac{p_1}{q_1} - \frac{a}{b}\right| \geqslant \frac{1}{bq_1},$$

so that

$$|b\alpha - a| > \frac{1}{q_1} = \frac{1}{a_1}.$$

On the other hand, it is obvious that

$$|1 \cdot \alpha - a_0| \leqslant \frac{1}{a_1},$$

so that

$$|b\alpha - a| > |1 \cdot \alpha - a_0| \qquad (1 \leqslant b),$$

which again contradicts the definition of a best approximation of the second kind. This proves Theorem 16.

Let us now consider the converse of Theorems 15 and 16. That the converse of Theorem 15 is false can be seen by considering $\frac{1}{2}$, which, as is easily shown, is an intermediate fraction for the number $\alpha = \frac{4}{5}$, while it is not a best approximation, since

$$\left|\frac{4}{5} - \frac{1}{1}\right| < \left|\frac{4}{5} - \frac{1}{2}\right| \qquad (1 < 2).$$

There are many more such examples, as the reader can verify for himself.

On the other hand, Theorem 16 does have an almost complete converse, which, of course, greatly enhances its value.

THEOREM 17. *Every convergent is a best approximation of the second kind, the sole exception being the trivial case of*

$$\alpha = a_0 + \frac{1}{2}, \qquad \frac{p_0}{q_0} = \frac{a_0}{1}.$$

PRELIMINARY REMARK. In the case of $\alpha = a_0 + \frac{1}{2}$, the fraction $p_0/q_0 = a_0/1$ is not a best approximation of the second kind because

$$|1 \cdot \alpha - (a_0 + 1)| = 1 |1 \cdot \alpha - a_0|.$$

PROOF. Let us examine the expression

$$|y\alpha - x|, \tag{29}$$

where y takes the values $1, 2, \cdots, q_k,$ and x can take arbitrary integral values. We denote by y_0 that value of y for which expression (29), after suitable choice of x, takes the smallest possible value. (If there are several such values of y, we take the *smallest* of these for y_0.) We denote by x_0 that value of x at which $|y_0\alpha - x|$ attains its minimum. It is easy to see that this value is unique. For if

$$\left|\alpha - \frac{x_0}{y_0}\right| = \left|\alpha - \frac{x_0'}{y_0}\right| \qquad (x_0 \neq x_0'),$$

we would have

$$\alpha = \frac{x_0 + x'_0}{2y_0}.$$

This fraction is irreducible. For if $x_0 + x'_0 = lp$ and $2y_0 = lq$ (with $l > 1$), we would have, for $l > 2$,

$$q < y_0, \quad \alpha = \frac{p}{q}, \quad |q\alpha - p| = 0,$$

which contradicts the definition of y_0; and for $l = 2$, we would have $q = y_0$ and

$$|q\alpha - p| = |y_0\alpha - p| = 0 < |y_0\alpha - x_0|,$$

which contradicts the definition of x_0.

Expanding the rational number α as a continued fraction, we thus obtain

$$\alpha = \frac{p_n}{q_n}, \quad p_n = x_0 + x'_0,$$

$$q_n = 2y_0 = a_n q_{n-1} + q_{n-2}, \quad a_n \geqslant 2,$$

so that if $a_n > 2$ or if $a_n = 2$ and $n > 1$, we have $q_{n-1} < y_0$. But

$$|q_{n-1}\alpha - p_{n-1}| = \frac{1}{q_n} = \frac{1}{2y_0} \leqslant \frac{1}{2} \leqslant |y_0\alpha - x_0|,$$

which contradicts the definition of y_0. If $n = 1$ and $a_n = 2$, we have $\alpha = a_0 + \frac{1}{2}$ and $y_0 = 1$, which is the one exceptional case.

Thus, the values y_0 and x_0 are uniquely defined by the given conditions. It directly follows from this that x_0/y_0 is a best approximation of the second kind for the number α, since the inequalities

$$|b\alpha - a| \leqslant |y_0\alpha - x_0|, \quad \frac{a}{b} \neq \frac{x_0}{y_0}, \quad b \leqslant y_0,$$

would obviously contradict the definitions of x_0 and y_0. From Theorem 16, we therefore have

$$x_0 = p_s, \quad y_0 = q_s \qquad (s \leqslant k).$$

If $s = k$, the theorem is proved. But if $s < k$, we obtain

$$|q_s\alpha - p_s| > \frac{1}{q_s + q_{s+1}} \geqslant \frac{1}{q_{k-1} + q_k}, \quad |q_k\alpha - p_k| \leqslant \frac{1}{q_{k+1}}, \quad -$$

and from the definitions of the numbers $p_s = x_0$ and $q_s = y_0$, we would have

$$|q_s \alpha - p_s| \leqslant |q_k \alpha - p_k|,$$

so that

$$\frac{1}{q_{k-1} + q_k} < \frac{1}{q_{k+1}},$$

that is,

$$q_{k+1} < q_k + q_{k-1},$$

which is impossible because of the rule by which the numbers q_k are formed. This completes the proof of Theorem 17.

Those properties of the apparatus of continued fractions that we have established in the present section were, historically, the original reason for the discovery and study of that apparatus. When Huygens set about constructing a model of the solar system by using toothed wheels, he was confronted with the problem of determining what numbers of teeth for the wheels would give a ratio for two interconnected wheels (equal to the ratio of their periods of rotation) that would be as close as possible to the ratio α of the periods of revolution of the corresponding planets. At the same time, the number of teeth obviously could not, for technical reasons, be too high. Thus, Huygens's problem was to find a rational number with numerator and denominator not exceeding a certain bound that would still be as close as possible to the given number α. (The number α might theoretically be irrational, but, in practice, it is assumed, in a given case, to be a rational fraction with very large numerator and denominator.) We have already seen that the theory of continued fractions provides the means of solving this problem.

7. The order of approximation

In the preceding section, we were concerned with evaluating the smallness of the difference $|\alpha - (p_k/q_k)|$ in comparison with other differences of the same type. Here, we shall make an absolute evaluation of this difference. Obviously, the only way of evaluating the smallness of $|\alpha - (p_k/q_k)|$ consists in comparing it with some decreasing function of q_k. To this end, Theorem 9 of Chapter I leads us directly to the inequality[3]

$$\left| \alpha - \frac{p_k}{q_k} \right| < \frac{1}{q_k^2}. \tag{30}$$

Therefore, the question must arise as to whether we can strengthen this inequality, that is, replace its right side with another function $f(q_k)$ of the denominator q_k that, for all $n \geq 1$, would satisfy the inequality

$$f(n) < \frac{1}{n^2}.$$

It is easy to see that if we want this strengthened form of inequality (30) to be satisfied for arbitrary α at all values of k, no significant strengthening in this direction is possible. More precisely, for any $\epsilon > 0$, we can always find a case for which

$$\left| \alpha - \frac{p_k}{q_k} \right| > \frac{1-\epsilon}{q_k^2}.$$

To show this, we need only examine the number

$$\alpha = [0; \; n, \; 1, \; n] = \frac{n+1}{n(n+2)},$$

for which,

$$p_1 = 1, \quad q_1 = n, \quad p_3 = n+1, \quad q_3 = n(n+2),$$

and therefore,

$$\left| \alpha - \frac{p_1}{q_1} \right| = \left| \frac{p_3}{q_3} - \frac{p_1}{q_1} \right| = \frac{1}{n(n+2)} = \frac{1}{q_1^2 \left(1 + \dfrac{2}{n} \right)}.$$

If we now choose n such that

$$\frac{1}{1 + \dfrac{2}{n}} > 1 - \epsilon,$$

we have

$$\left| \alpha - \frac{p_1}{q_1} \right| > \frac{1-\epsilon}{q_1^2}.$$

---

[3] If $\alpha = p_k/q_k$ (when Theor. 9 is inapplicable because there is no q_{k+1}), inequality (30) becomes trivial.

However, if we relax the requirement that the strengthened inequality be satisfied for arbitrary α at all values of k (without exception), we can then obtain a number of interesting and important propositions, as we proceed to demonstrate.

THEOREM 18. *If a number α has a convergent of order $k > 0$, at least one of the following two inequalities must hold:*

$$\left| \alpha - \frac{p_k}{q_k} \right| < \frac{1}{2q_k^2}, \quad \left| \alpha - \frac{p_{k-1}}{q_{k-1}} \right| < \frac{1}{2q_{k-1}^2}.$$

PROOF. Since α lies between p_{k-1}/q_{k-1} and p_k/q_k, we have

$$\left| \alpha - \frac{p_k}{q_k} \right| + \left| \alpha - \frac{p_{k-1}}{q_{k-1}} \right| = \left| \frac{p_k}{q_k} - \frac{p_{k-1}}{q_{k-1}} \right| = \frac{1}{q_k q_{k+1}} < \frac{1}{2q_k^2} + \frac{1}{2q_{k-1}^2}.$$

(The inequality expresses the fact that the geometric mean of the quantities $1/q_k^2$ and $1/q_{k-1}^2$ is less than their arithmetic mean; equality would be possible only if $q_k = q_{k-1}$, which in the present case is ruled out.) The assertion of the theorem follows immediately.

This proposition is interesting because it has a converse (in a certain sense).

THEOREM 19. *Every irreducible rational fraction a/b that satisfies the inequality*

$$\left| \alpha - \frac{a}{b} \right| < \frac{1}{2b^2}$$

is a convergent of the number α.

PROOF. On the basis of Theorem 16, it is sufficient to show that the fraction a/b is a best approximation of the second kind of the number α. Suppose that

$$| d\alpha - c | \leqslant | b\alpha - a | < \frac{1}{2b} \qquad \left(d > 0, \ \frac{c}{d} \neq \frac{a}{b} \right);$$

then,

$$\left| \alpha - \frac{c}{d} \right| < \frac{1}{2bd}$$

and, consequently,

$$\left| \frac{c}{d} - \frac{a}{b} \right| \leqslant \left| \alpha - \frac{c}{d} \right| + \left| \alpha - \frac{a}{b} \right| < \frac{1}{2bd} + \frac{1}{2b^2} = \frac{b+d}{2b^2 d}. \qquad (31)$$

On the other hand, since $c/d \neq a/b$, we have

$$\left| \frac{c}{d} - \frac{a}{b} \right| \geqslant \frac{1}{bd}.$$

Therefore, inequality (31) implies

$$\frac{1}{bd} < \frac{b+d}{2b^2d},$$

so that $d > b$. Thus, the fraction a/b is a best approximation of the second kind of the number α and Theorem 19 is proved.

A further strengthening of Theorem 18 is the following, considerably more profound theorem.

THEOREM 20.[4] *If a number α has a convergent of order $k > 1$, at least one of the following three inequalities must hold:*

$$\left| \alpha - \frac{p_k}{q_k} \right| < \frac{1}{\sqrt{5}q_k^2}, \qquad \left| \alpha - \frac{p_{k-1}}{q_{k-1}} \right| < \frac{1}{\sqrt{5}q_{k-1}^2},$$

$$\left| \alpha - \frac{p_{k-2}}{q_{k-2}} \right| < \frac{1}{\sqrt{5}q_{k-2}^2}.$$

PROOF. Let us define, for $k \geq 1$,

$$\frac{q_{k-2}}{q_{k-1}} = \varphi_k, \qquad \varphi_k + r_k = \psi_k.$$

LEMMA. *If $k \geq 2$, $\psi_k \leq \sqrt{5}$, and $\psi_{k-1} \leq \sqrt{5}$, then*

$$\varphi_k > \frac{\sqrt{5}-1}{2}.$$

PROOF. Since

$$\frac{1}{\varphi_{n+1}} = \frac{q_n}{q_{n-1}} = a_n + \varphi_n \tag{32}$$

and

$$r_n = a_n + \frac{1}{r_{n+1}},$$

it follows that

$$\frac{1}{\varphi_{n+1}} + \frac{1}{r_{n+1}} = \varphi_n + r_n = \psi_n,$$

and, from the conditions of the lemma,

$$\varphi_k + r_k \leq \sqrt{5}, \qquad \frac{1}{\varphi_k} + \frac{1}{r_k} \leq \sqrt{5},$$

---

[4] Some simplification of the proof given here appears in the article by I. I. Zhogin, "Variant dokazatel'stva odnoi teoremy iz teorii tsepnykh drobei" ("A variation of the proof of a theorem in the theory of continued fractions"), *Uspekhi matematicheskikh nauk*, **12**, No. 3, 321–322 (1957).

so that

$$\left(\sqrt{5} - \varphi_k\right)\left(\sqrt{5} - \frac{1}{\varphi_k}\right) \geqslant 1,$$

or, since φ_k is a rational number,

$$5 - \sqrt{5}\left(\varphi_k + \frac{1}{\varphi_k}\right) > 0.$$

Then, since $\varphi_k > 0$, we obtain

$$\left(\frac{\sqrt{5}}{2} - \varphi_k\right)^2 < \frac{1}{4},$$

and, consequently,

$$\frac{\sqrt{5}}{2} - \varphi_k < \frac{1}{2}, \qquad \varphi_k > \frac{\sqrt{5}-1}{2},$$

which proves the lemma.

Let us now suppose that, in contradiction to our assertion,

$$\left|\alpha - \frac{p_n}{q_n}\right| \geqslant \frac{1}{\sqrt{5}q_n^2} \qquad (n = k,\ k-1,\ k-2).$$

From formula (16) of Chapter I, we have

$$\left|\alpha - \frac{p_n}{q_n}\right| = \left|\frac{p_n r_{n+1} + p_{n-1}}{q_n r_{n+1} + q_{n-1}} - \frac{p_n}{q_n}\right|$$

$$= \frac{1}{q_n(q_n r_{n+1} + q_{n-1})} = \frac{1}{q_n^2(r_{n+1} + \varphi_{n+1})} = \frac{1}{q_n^2 \psi_{n+1}},$$

and, consequently,

$$\psi_{n+1} \leqslant \sqrt{5} \qquad (n = k,\ k-1,\ k-2).$$

We conclude, on the basis of our lemma, that

$$\varphi_k > \frac{\sqrt{5}-1}{2}, \qquad \varphi_{k+1} > \frac{\sqrt{5}-1}{2},$$

and hence, because of (32),

$$a_k = \frac{1}{\varphi_{k+1}} - \varphi_k < \frac{2}{\sqrt{5}-1} - \frac{\sqrt{5}-1}{2} = 1,$$

which is impossible. This contradiction completes the proof of Theorem 20.

Theorems 18 and 20 give the obvious impression of the beginning of a series of propositions that will admit yet further extension. However, this impression is erroneous. Consider the number

$$\alpha = [1; \ 1, \ 1, \ \ldots].$$

Assuming, as usual, that $\alpha = 1 + (1/r_1)$, we obviously have $r_1 = \alpha$, so that

$$\alpha = 1 + \frac{1}{\alpha}, \qquad \alpha^2 - \alpha - 1 = 0,$$

and, consequently,

$$\alpha = \frac{1 + \sqrt{5}}{2}.$$

Since, obviously, $r_n = \alpha$ for arbitrary n, we have

$$\alpha = \frac{p_k \alpha + p_{k-1}}{q_k \alpha + q_{k-1}},$$

and, consequently,

$$\left| \alpha - \frac{p_k}{q_k} \right| = \frac{1}{q_k(q_k \alpha + q_{k-1})} = \frac{1}{q_k^2 \left(\alpha + \frac{q_{k-1}}{q_k} \right)}.$$

But from Theorem 6 of Chapter I, we have

$$\frac{q_k}{q_{k-1}} = [1; \ 1, \ 1, \ \ldots, \ 1] \to \alpha \quad \text{as} \quad k \to \infty$$

so that

$$\frac{q_{k-1}}{q_k} = \frac{1}{\alpha} + \varepsilon_k = \frac{\sqrt{5}-1}{2} + \varepsilon_k \qquad (\varepsilon_k \to 0 \quad \text{as} \quad k \to \infty).$$

Thus,

$$\left| \alpha - \frac{p_k}{q_k} \right| = \frac{1}{q_k^2 \left(\frac{\sqrt{5}+1}{2} + \frac{\sqrt{5}-1}{2} + \varepsilon_k \right)} = \frac{1}{q_k^2 (\sqrt{5} + \varepsilon_k)}.$$

This shows that, no matter what the number $c < (1/\sqrt{5})$ may be, for sufficiently large k, we will obtain

$$\left| \alpha - \frac{p_k}{q_k} \right| > \frac{c}{q_k^2}.$$

Thus, the constant $1/\sqrt{5}$ in Theorem 20 cannot be replaced by any smaller constant if we wish the corresponding inequality to be satisfied for an infinite set of values of k with arbitrary α. For every smaller constant, there exists an α [namely, $\alpha = \frac{1}{2}(\sqrt{5} + 1)$] that can satisfy the required inequality for no more than a finite number of values of k. Thus, the chain of propositions that begins with Theorems 18 and 20 is broken after the latter theorem, and admits no further continuation.

8. *General approximation theorems*

Up to now, we have been primarily interested in approximations given by convergents and have clarified a number of fundamental questions associated with this problem. Since we have seen that the convergents are best approximations, we may assume that the obtained results will allow us to study, in full measure, the rules that govern the approximation of irrational numbers by rational fractions, independently of any particular representing apparatus. We now turn to problems of this type. It is, of course, impossible (within the framework of the present elementary monograph) to give any sort of complete exposition of the fundamentals of the corresponding theory, partly because of lack of space, but primarily because such an exposition would have only an indirect bearing on our problem. We shall confine ourselves to presenting a number of elementary propositions, which will illustrate the application of continued fractions to the study of the arithmetic nature of irrational numbers.

The first problem that naturally arises in connection with the results of the preceding section may be formulated as follows: For what constants c does the inequality

$$\left| \alpha - \frac{p}{q} \right| < \frac{c}{q^2} \tag{33}$$

have an infinite set of solutions in integers p and q, $q > 0$, for arbitrary real α? The final result of the preceding section leads us to the following theorem.

THEOREM 21. *Inequality (33) has an infinite set of solutions in integers p and q ($q > 0$) for arbitrary real α if $c \geq (1/\sqrt{5})$. However, if*

$c < (1/\sqrt{5})$, *inequality (33) will, for suitably chosen* α, *have only a finite number of such solutions.*

The first assertion is an immediate consequence of Theorem 20. (In the case in which α is a rational number a/b and, therefore, has only a finite number of convergents, the first assertion of Theorem 21 can be proved in a trivial manner by setting $q = nb$ and $p = na$, for $n = 1$, $2, 3, \cdots$). Suppose, then, that $c < (1/\sqrt{5})$. As in section 7, let us set

$$\alpha = \frac{1+\sqrt{5}}{2} = [1;\ 1,\ 1,\ \ldots].$$

If two integers p and q $(q > 0)$ satisfy inequality (33), Theorem 19 tells us that p/q is a convergent of the number α. But we saw at the end of section 7 that only a finite number of these convergents satisfy inequality (33) under our hypothesis that $c < (1/\sqrt{5})$. This proves our assertion.

Thus, in general (that is, if we consider all possible real numbers α), the order of approximation characterized by the quantity $1/(\sqrt{5}q^2)$ cannot be improved. (The term "order of approximation" refers to that magnitude of error within which a suitable estimate can always be found.)[5] This does not mean that there are no individual irrational numbers for which approximations of much higher order are possible. On the contrary, the possibilities in this direction are boundless—a fact that is most easily shown by the apparatus of continued fractions.

THEOREM 22. *For any positive function* $\varphi(q)$ *with natural argument* q, *there is an irrational number* α *such that the inequality*

$$\left| \alpha - \frac{p}{q} \right| < \varphi(q)$$

has an infinite number of solutions in integers p *and* q $(q > 0)$.

PROOF. Let us construct an infinite continued fraction α by choosing its elements successively in such a way that they will satisfy the inequalities

$$a_{k+1} > \frac{1}{q_k^2 \varphi(q_k)} \qquad (k \geqslant 0).$$

This, of course, can be done in an infinite number of ways. Here, a_0 can be chosen arbitrarily. Then, for any $k \geq 0$,

$$\left| \alpha - \frac{p_k}{q_k} \right| < \frac{1}{q_k q_{k+1}} = \frac{1}{q_k(a_{k+1}q_k + q_{k-1})} \leqslant \frac{1}{a_{k+1}q_k^2} < \varphi(q_k),$$

which proves the theorem.

[5] Translation editor's note.

We now note that, in the most general case, the inequalities

$$\frac{1}{q_k(q_k + q_{k+1})} < \left| \alpha - \frac{p_k}{q_k} \right| \leqslant \frac{1}{q_k q_{k+1}}$$

or, equivalently,

$$\frac{1}{q_k^2 \left(a_{k+1} + 1 + \dfrac{q_{k-1}}{q_k} \right)} < \left| \alpha - \frac{p_k}{q_k} \right| \leqslant \frac{1}{q_k^2 a_{k+1} + \dfrac{q_{k-1}}{q_k}}$$

imply

$$\frac{1}{q_k(a_{k+1} + 2)} < \left| \alpha - \frac{p_k}{q_k} \right| \leqslant \frac{1}{q_k^2 a_{k+1}}, \qquad (34)$$

from which it is clear that, for given $a_0, a_1, \cdots, a_k$, the greater the subsequent element a_{k+1} is, the more closely the fraction p_k/q_k will approximate the number α. And since the convergents are, in all cases, best approximations, we arrive at the conclusion that those irrational numbers whose elements include large numbers admit good approximation by rational fractions. This qualitative remark is expressed quantitatively in inequality (34). In particular, irrational numbers with bounded elements admit the worst approximations. Thus, it becomes clear why we have repeatedly chosen the number

$$\frac{\sqrt{5} + 1}{2} = [1; 1, 1, \ldots]$$

when we wished to exhibit an irrational number that did not admit approximations of higher than a fixed order. Of all irrational numbers, this clearly has the smallest possible elements (excluding a_0, which plays no role here) and hence is the most poorly approximated by rational fractions.

Those approximating properties that are peculiar to numbers with bounded elements are completely expressed in the following proposition, which, after what has already been said, is almost obvious.

THEOREM 23. *For every irrational number α with bounded elements, and for sufficiently small c, the inequality*

$$\left| \alpha - \frac{p}{q} \right| < \frac{c}{q^2}$$

has no solution in integers p and q ($q > 0$). On the other hand, for every number α with an unbounded sequence of elements and arbitrary $c > 0$, inequality (33) has an infinite set of such solutions.

In other words, irrational numbers with bounded elements admit an order of approximation no higher than $1/q^2$, while every irrational number with unbounded elements admits a higher order of approximation.

PROOF. If the set of elements of the continued fraction representing α is not bounded above, then for arbitrary positive c there is an infinite set of integers k such that

$$a_{k+1} > \frac{1}{c},$$

and, consequently, on the basis of the second of the inequalities in (34), there is an infinite set of integers k such that

$$\left| \alpha - \frac{p_k}{q_k} \right| < \frac{c}{q_k^2},$$

which proves the second assertion of the theorem. If there exists an $M > 0$ such that

$$a_k < M \qquad (k = 1, 2, \ldots),$$

then, on the basis of the first of the inequalities in (34), we have, for arbitrary $k \geq 0$,

$$\left| \alpha - \frac{p_k}{q_k} \right| > \frac{1}{q_k^2 (M+2)}.$$

Now let p and q be arbitrary integers $(q > 0)$, and let k be determined by the inequalities

$$q_{k-1} < q \leq q_k.$$

Then, since all convergents are best approximations of the first kind,

$$\left| \alpha - \frac{p}{q} \right| \geq \left| \alpha - \frac{p_k}{q_k} \right| > \frac{1}{q_k^2 (M+2)}$$

$$= \frac{1}{q^2 (M+2)} \left(\frac{q}{q_k} \right)^2 > \frac{1}{q^2 (M+2)} \left(\frac{q_{k-1}}{q_k} \right)^2$$

$$= \frac{1}{q^2 (M+2)} \left(\frac{q_{k-1}}{a_k q_{k-1} + q_{k-2}} \right)^2$$

$$> \frac{1}{q^2 (M+2)} \frac{1}{(a_k + 1)^2} > \frac{1}{(M+2)(M+1)^2 q^2}.$$

Thus, if we choose

$$c < \frac{1}{(M+2)(M+1)^2},$$

inequality (33) cannot be satisfied for any pair of integers p and q ($q > 0$). This proves the first assertion of the theorem.

Up to this point, we have always evaluated the closeness of an approximation in terms of the smallness of the difference $\alpha - (p/q)$; however, we might have considered instead the difference $q\alpha - p$ (as in sec. 6), making the appropriate changes in the formulation of all the theorems. This simple observation leads directly to a certain new and extremely important aspect of the problem that we are studying.

The simplest homogeneous linear equation with two unknowns x and y, namely,

$$\alpha x - y = 0, \tag{35}$$

where α is a given irrational number, obviously cannot be exactly solved in whole numbers (except, of course, in the trivial case of $x = y = 0$). However, we may pose the problem of obtaining an approximate solution, that is, of choosing integers x and y for which the difference $\alpha x - y$ is sufficiently small (that is, less than a preassigned amount). Obviously, all the preceding theorems of this section can be interpreted as confirmation of the rules governing this kind of approximate solution to equation (35) in whole numbers. Thus, for example, Theorem 21 shows that there always exists an infinite set of pairs of integers x and y ($x > 0$), such that

$$|\alpha x - y| < \frac{C}{x}, \tag{36}$$

for any positive C greater than or equal to $1/\sqrt{5}$.

With this approach, it is natural to pass from the homogeneous equation (35) to the non-homogeneous equation

$$\alpha x - y = \beta \tag{37}$$

(where β is a given real number) and to investigate the existence and nature of its approximate solutions in integers x and y (in other words, to investigate the principles involved in attempting to make the difference $\alpha x - y - \beta$ as small as possible by a suitable choice of integers x and y). This problem was first posed by the great Russian mathematician P. L. Chebyshev, who obtained the first basic results connected with it, and has been the subject of continued intensive study, especially by the Soviet arithmetic school.

The first basic feature distinguishing the non-homogeneous case from the homogeneous one is that it is possible to make the quantity $|ax - y - \beta|$ arbitrarily small for arbitrary β by a suitable choice of x and y only if the number a is irrational (whereas, in the homogeneous case, the quantity $|ax - y|$ can be made arbitrarily small for *any* a). In fact, if $a = a/b$, where $b > 0$ and a are integers, then, by setting $\beta = 1/2b$, we obtain, for arbitrary integers x and y,

$$|ax - y - \beta| = \left| \frac{2(ax - by) - 1}{2b} \right| \geqslant \frac{1}{2b},$$

since $|2(ax - by) - 1|$, being an odd integer, is at least equal to unity.

Thus, in all that follows, we shall assume a to be irrational. With this understanding, we shall now show that not only is it also possible to make the quantity $|ax - y - \beta|$ arbitrarily small, but the analogy with the homogeneous case can be extended considerably further.

THEOREM 24 (Chebyshev). *For an arbitrary irrational number a and an arbitrary real number β, the inequality $|ax - y - \beta| < 3/x$ has an infinite set of solutions in integers x and y (where $x > 0$).*[6]

PRELIMINARY REMARK. Obviously, this result is completely analogous to the corresponding problem for homogeneous equations, expressed in Theorem 21. The difference consists only in the fact that here, instead of $1/\sqrt{5}$, we have 3. The order of the approximation is the same as before. We note also that the number 3 is not the best possible and that the exact value of the greatest lower bound of the set of numbers that would verify Theorem 24 is considerably less than 3.

PROOF. Let p/q be an arbitrary convergent of a. We then have

$$a = \frac{p}{q} + \frac{\delta}{q^2} \qquad (0 < |\delta| < 1); \tag{38}$$

also, for any real β, we can find a number t such that

$$|q\beta - t| \leqslant \frac{1}{2},$$

---

[6] A simple proof of a somewhat stronger theorem is found in Khinchin's article, "Printsip Dirikhle v teorii diofantovykh priblizhenii" ("Dirichlet's principle in the theory of Diophantine approximations"), *Uspekhi matematicheskikh nauk*, **3**, No. 3, 17–18 (1948). Further refinements are contained in Khinchin's article, "O zadache Chebysheva" ("On a problem of Chebyshev"), *Izvestiya akad. nauk SSSR, ser. matem.*, **10**, 281–294 (1946). (B. G.)

so that

$$\beta = \frac{t}{q} + \frac{\delta'}{2q} \qquad (|\delta'| \leqslant 1). \tag{39}$$

Since p and q have no common divisors other than ± 1, there exists a pair of integers x and y such that

$$\frac{q}{2} \leqslant x < \frac{3q}{2}, \qquad px - qy = t.$$

For if r/s is the convergent immediately preceding p/q,

$$qr - ps = \varepsilon = \pm 1, \qquad q(\varepsilon rt) - p(\varepsilon st) = t\varepsilon^2 = t,$$

and for an arbitrary integer k,

$$p(kq - \varepsilon st) - q(kp - \varepsilon rt) = t;$$

but k can be chosen so that

$$\frac{q}{2} \leqslant x = kq - \varepsilon st < \frac{3q}{2}.$$

Then, on the basis of equations (38) and (39),

$$|ax - y - \beta| = \left| \frac{xp}{q} + \frac{x\delta}{q^2} - y - \frac{t}{q} - \frac{\delta'}{2q} \right|$$

$$= \left| \frac{x\delta}{q^2} - \frac{\delta'}{2q} \right| < \frac{x}{q^2} + \frac{1}{2q},$$

and since

$$q > \frac{2}{3} x,$$

we have

$$|ax - y - \beta| < \frac{9}{4x} + \frac{3}{4x} = \frac{3}{x}.$$

Finally, since q can be chosen arbitrary large and since $x \geq q/2$, it follows that x can be arbitrarily large. This proves the theorem.

But the problem of an approximate solution to equation (37) in whole numbers can be put in a different, and somewhat more natural, form. Since the crux of the problem is to make the quantity $|ax - y - \beta|$ as small as possible with as large integral values of x and y as possible, it is most natural to state the problem in the following manner. We know (from Theor. 24) that, for any positive number n (no matter how large), any irrational number α, and any real β, we can find integers $x > 0$ and y satisfying the inequality

$$|\alpha x - y - \beta| < \frac{1}{n}. \tag{40}$$

However, Theorem 24 does not generally give us any information as to the limits within which we should seek these numbers so as to attain the required accuracy, characterized by the quantity $1/n$. This might be achieved, for example, if we could exhibit some number N, dependent on n, but independent of α and β, such that inequality (40) would always be satisfied under the additional condition that

$$|x| \leqslant N.$$

This new statement of the problem is obviously quite different from the original one. Whereas formerly (as in Theor. 24) the accuracy of approximation was determined by the value of x, we now wish to fix this accuracy in advance and see how large a value of x we should choose to attain this accuracy. The solution to the problem is significantly altered by this difference in its statement. Specifically, we obtain quite different results in the homogeneous and non-homogeneous cases.

In the case of a homogeneous equation ($\beta = 0$), the stated problem has a very simple solution.

THEOREM 25. *For all real numbers $n \geq 1$ and α, there are integers x and y satisfying the inequalities*

$$0 < x \leqslant n, \qquad |\alpha x - y| < \frac{1}{n}.$$

PROOF. If α is a rational number, a/b, such that $0 < b \leq n$, the conclusion is immediate for $x = b$ and $y = a$. If α either is irrational or has a denominator exceeding n, we define k by the relationship

$$q_k \leqslant n < q_{k+1}$$

(where p_k/q_k denotes the kth-order convergent of α) and obtain

$$\left| \alpha - \frac{p_k}{q_k} \right| \leqslant \frac{1}{q_k q_{k+1}} < \frac{1}{q_k n},$$

so that

$$|\alpha q_k - p_k| < \frac{1}{n}, \qquad 0 < q_k \leqslant n,$$

which proves the theorem.

Now, we naturally ask whether we may obtain the same order of approximation in the case of the non-homogeneous equation (37). In other words, may we assert that, for any irrational number α, a positive number C can be found such that, for any $n \geq 1$ and β, there exist integers x and y satisfying the inequalities

$$0 < x \leqslant Cn, \qquad |\alpha x - y - \beta| < \frac{1}{n}?$$

(Clearly, we are now asking even less than for the homogeneous case, since we are allowing C to depend on α, whereas in the homogeneous case, $C = 1$ was an absolute constant.) It is easy to give certain a priori arguments against the possibility of such an assumption. First, for rational α, it is clearly untrue, since, as we have seen, the quantity $|\alpha x - y - \beta|$ cannot in general (that is, with arbitrary β) be made arbitrarily small. This leads us to expect that if α is irrational (but is approximated extremely closely by rational fractions), the quantity $|\alpha x - y - \beta|$, even though it can (on the basis of Theor. 24) be made arbitrarily small, requires comparatively large values of x and y to accomplish this with a suitably chosen value of β. These considerations also lead us to suppose that the more poorly the number α is approximated by rational fractions (that is, the more difficult it is to make the quantity $\alpha x - y$ approach zero), the easier it will be to have the difference $\alpha x - y$ approach an arbitrary real number β. As we know, this, in turn, requires that the elements of the number α not increase too fast. All these preliminary considerations are expressed precisely in the following theorem.

THEOREM 26. *For the existence of a positive number C with the property that, for arbitrary real numbers $n \geq 1$ and β, two integers x and y ($x > 0$) exist satisfying the inequalities*

$$x \leqslant Cn, \qquad |\alpha x - y - \beta| < \frac{1}{n},$$

it is necessary and sufficient that the irrational number α be represented by a continued fraction with bounded elements.

PROOF. Suppose that $\alpha = [a_0; a_1, a_2, \cdots]$, that $a_i < M$ (for $i = 1, 2, \cdots$), that $m \geq 1$, and that β is an arbitrary real number. Denoting by p_k/q_k the convergents of the number α, we can determine the subscript k from the inequalities

$$q_k \leqslant m < q_{k+1};$$

then,

$$\left|\alpha - \frac{p_k}{q_k}\right| < \frac{1}{q_k q_{k+1}} < \frac{1}{mq_k},$$

or

$$\alpha = \frac{p_k}{q_k} + \frac{\delta}{mq_k} \qquad (|\delta| \leqslant 1). \tag{41}$$

We now choose an integer t such that

$$|\beta q_k - t| \leqslant \frac{1}{2},$$

so that

$$\beta = \frac{t}{q_k} + \frac{\delta'}{2q_k} \qquad (|\delta'| \leqslant 1). \tag{42}$$

Finally, as in the proof of Theorem 24, we find a pair of integers x and y satisfying the relationships

$$xp_k - yq_k = t, \qquad 0 < x \leqslant q_k. \tag{43}$$

It follows from (41), (42), and (43) that

$$|\alpha x - y - \beta| = \left| \frac{xp_k}{q_k} - y - \frac{t}{q_k} + \frac{x\delta}{mq_k} - \frac{\delta'}{2q_k} \right|$$

$$= \left| \frac{x\delta}{mq_k} - \frac{\delta'}{2q_k} \right| < \frac{x}{mq_k} + \frac{1}{2q_k} \leqslant \frac{1}{m} + \frac{1}{2q_{k+1}}\left(\frac{q_{k+1}}{q_k}\right)$$

$$< \frac{1}{m} + \frac{1}{2m}(a_{k+1} + 1) < \frac{1}{m} + \frac{M+1}{2m} = \frac{M+3}{2m}.$$

Up to now, the number $m \geq 1$ has been completely arbitrary. If we now set $m = \frac{1}{2}(M+3)n$ for given $n \geq 1$, we shall obviously have $m > 1$. Consequently, from what was stated above, if we choose the numbers x and y as indicated, we have

$$0 < x \leqslant q_k \leqslant m = \frac{M+3}{2}n,$$

$$|\alpha x - y - \beta| < \frac{1}{n},$$

which proves the first part of the theorem.

To prove the second part, let us suppose that the set of elements a_k of the number α is unbounded above. Theorem 23 than indicates that, in this case, for any positive number ϵ there are integers $q > 0$ and p satisfying the inequality

$$\left| \alpha - \frac{p}{q} \right| < \frac{\epsilon^2}{q^2},$$

so that

$$\alpha = \frac{p}{q} + \frac{\delta\epsilon^2}{q^2} \qquad (|\delta| < 1).$$

We now set $n = q/\epsilon$ and $\beta = 1/2q$. Then, for arbitrary integers x and y (with $0 < x \leq Cn$), we obtain

$$|\alpha x - y - \beta| = \left| \frac{xp}{q} - y - \frac{1}{2q} + \frac{x\delta\epsilon^2}{q^2} \right| = \left| \frac{2(xp-yq)-1}{2q} + \frac{x\delta\epsilon^2}{q^2} \right|$$

$$> \frac{2(xp-yq)-1|}{2q} - \frac{x\epsilon^2}{q^2} \geq \frac{1}{2q} - \frac{C\epsilon}{q} = \frac{1-2C\epsilon}{2q} = \frac{1-2C\epsilon}{2\epsilon} \cdot \frac{1}{n}.$$

But no matter how large C may be, for sufficiently small ϵ, we shall have $[(1 - 2C\epsilon)/2\epsilon] > 1$, and, consequently, for arbitrary integers x and y (with $0 < x \leq Cn$), we obtain

$$|\alpha x - y - \beta| > \frac{1}{n},$$

which proves the first part of the theorem.

Let us review the results that we have obtained. In investigating the approximate solutions to equation (37) in whole numbers, we must examine as a "normal" case the one in which the accuracy characterized by the quantity $1/n$ can be attained for arbitrary $n \geq 1$ at some $x < Cn$, where C is a constant (possibly depending on α). A homogeneous equation (obtained for $\beta = 0$) always has a normal solution (Theor. 25). Theorem 26 shows that the general (non-homogeneous) equation has a normal solution if, and only if, the corresponding homogeneous equation has no "supernormal" solution (that is, if it is impossible to satisfy the homogeneous equation with integers $x > 0$ and y such that $x < \epsilon n$ for arbitrary $\epsilon > 0$ and properly chosen n, with an accuracy of $1/n$). From this point of view, the results of our investigation can be regarded as a variation of the general law concerning the solution of linear equations (algebraic, integral, etc.): *In the general case, a non-homogeneous equation can be solved "normally" if the corresponding homogeneous equation has no "supernormal" solution.*

We note also that in Theorem 26 we required that C be independent of β. The same result would hold if we allowed C to be a function of β, but the proof (second part) would be somewhat more complicated.

9. The approximation of algebraic irrational numbers and Liouville's transcendental numbers

Suppose that

$$f(x) = a_0 + a_1 x + \ldots + a_n x^n \tag{44}$$

is a polynomial of degree n with *integral* coefficients $a_0, a_1, \cdots, a_n$. Then, a root, α, of this polynomial is said to be *algebraic*. Since every rational number $\alpha = a/b$ can be defined as the root of the first-degree equation $bx - a = 0$, the concept of an algebraic number is clearly a natural generalization of the concept of a rational number. If a given algebraic number satisfies an equation $f(x) = 0$ of degree n, and does not satisfy any equation of lower degree (with integral coefficients), it is called an algebraic number of degree n. In particular, rational numbers can be defined as first-degree algebraic numbers. The number $\sqrt{2}$, being a root of the polynomial $x^2 - 2$, is a second-degree algebraic number, or, as we say, a quadratic irrational. Cubic, fourth-degree, and higher irrationals are defined analogously. All non-algebraic numbers are said to be *transcendental*. Examples of transcendental numbers are e and π. Because of the great role that algebraic numbers play in contemporary number theory, many special studies have been devoted to the question of their properties with regard to their approximation by rational fractions. The first noteworthy result in this direction was the following theorem, known as Liouville's theorem.

THEOREM 27. *For every real irrational algebraic number α of degree n, there exists a positive number C such that, for arbitrary integers p and q $(q > 0)$,*

$$\left| \alpha - \frac{p}{q} \right| > \frac{C}{q^n}.$$

PROOF. Suppose that α is a root of the polynomial (44). From algebra, we may write

$$f(x) = (x - \alpha) f_1(x), \tag{45}$$

where $f_1(x)$ is a polynomial of degree $n - 1$. Here, $f_1(\alpha) \neq 0$. To show this, suppose that $f_1(\alpha) = 0$. Then, the polynomial $f_1(x)$ could be divided (without a remainder) by $x - \alpha$ and, hence, the polynomial $f(x)$ could be divided by $(x - \alpha)^2$. But, then, the derivative $f'(x)$ could be divided by $x - \alpha$; that is, we would have $f'(\alpha) = 0$, which is impossible since $f'(x)$ is a polynomial of degree $n - 1$ with integral coefficients and a is an algebraic number of degree n. Hence, $f_1(a) \neq 0$, and, consequently, we can find a positive number δ such that

$$f_1(x) \neq 0 \qquad (a - \delta \leqslant x \leqslant a + \delta).$$

Suppose that p and q $(q > 0)$, are an arbitrary pair of integers. If $|a - (p/q)| \leq \delta$, then $f_1(p/q) \neq 0$, and, by substituting $x = p/q$ in identity (45), we obtain

$$\frac{p}{q} - a = \frac{f\left(\frac{p}{q}\right)}{f_1\left(\frac{p}{q}\right)} = \frac{a_0 + a_1\left(\frac{p}{q}\right) + \ldots + a_n\left(\frac{p}{q}\right)^n}{f_1\left(\frac{p}{q}\right)}$$

$$= \frac{a_0 q^n + a_1 p q^{n-1} + \ldots + a_n p^n}{q^n f_1\left(\frac{p}{q}\right)}.$$

The numerator of this fraction is an integer. It is also non-zero, because otherwise we would have $a = p/q$, whereas a is by hypothesis irrational. Consequently, this numerator is at least equal to unity in absolute value. We denote by M the least upper bound of the function $f_1(x)$ in the interval $(a - \delta, a + \delta)$. From the last inequality, we thus obtain

$$\left| a - \frac{p}{q} \right| \geqslant \frac{1}{Mq^n}.$$

In the event that

$$\left| a - \frac{p}{q} \right| > \delta,$$

it follows that

$$\left| a - \frac{p}{q} \right| > \frac{\delta}{q^n};$$

and if we now denote by C any positive number less than δ and $1/M$, we obtain, in both cases (that is, for arbitrary $q > 0$ and p),

$$\left| a - \frac{p}{q} \right| > \frac{C}{q^n},$$

which completes the proof of Theorem 27.

Liouville's theorem shows that algebraic numbers do not admit rational-fraction approximations of greater than a certain order of accuracy (this depending basically on the degree of the algebraic number in question). The main historical importance of this theorem consisted in the fact that it made possible the proof of the existence of transcendental numbers, and enabled one to give specific examples of such numbers. As we have seen, to do this, it is sufficient to exhibit an irrational number for which rational fractions give extremely close approximations, and theorem 22 shows that the possibilities for this are unlimited.

Specifically, Theorem 27 shows that if for arbitrary $C > 0$ and arbitrary natural n there exist integers p and $q(q > 0)$, such that

$$\left| \alpha - \frac{p}{q} \right| \leqslant \frac{C}{q^n},$$ (46)

then the number α is transcendental. Using the apparatus of continued fractions, it is very easy to exhibit as many such numbers as we desire. All that is necessary is to choose elements $a_0, a_1, \cdots , a_k$, form the convergent p_k/q_k, and take

$$a_{k+1} > q_k^{k-1},$$

since then

$$\left| \alpha - \frac{p_k}{q_k} \right| < \frac{1}{q_k q_{k+1}} < \frac{1}{q_k^2 a_{k+1}} < \frac{1}{q_k^{k+1}}.$$

As a result of the above, inequality (46) is obviously satisfied for sufficiently large values of k, no matter what $C > 0$ and natural n may be.

10. Quadratic irrational numbers and periodic continued fractions

Theorem 27 shows that, for any quadratic irrational number α, there exists a positive number C, depending on α, such that the inequality

$$\left| \alpha - \frac{p}{q} \right| < \frac{C}{q^2}$$

has no solution in integers p and $q(q > 0)$. From this and from Theorem 23, it follows that the elements of every quadratic irrational number are bounded. Long before Liouville, however, Lagrange had discovered a much more significant property of the continued fractions representing these irrationals (one that is even more characteristic of them). It turns out that a sequence of quadratic irrational elements is always a periodic sequence and, conversely, that every periodic continued fraction represents some quadratic irrational number. The present section is devoted to a proof of this assertion.

Let us agree to call the continued fraction

$$\alpha = [a_0; \ a_1, \ a_2, \ \ldots]$$

periodic if there exist positive integers k_0 and h such that, for arbitrary $k \geq k_0$,

$$a_{k+h} = a_k.$$

In analogy with the procedure for decimal fractions, we shall indicate such a periodic continued fraction as follows:

$$\alpha = [a_0; \ a_1, \ a_2, \ \ldots, \ a_{k_0-1}, \ \overline{a_{k_0}, \ a_{k_0+1}, \ \ldots, \ a_{k_0+h-1}}]. \quad (47)$$

THEOREM 28. *Every periodic continued fraction represents a quadratic irrational number and every quadratic irrational number is represented by a periodic continued fraction.*

PROOF. The first assertion can be proved in a few words. Obviously, the remainders of the periodic continued fraction (47) satisfy the relationship

$$r_{k+h} = r_k \qquad (k \gg k_0).$$

Therefore, on the basis of formula (16) of Chapter I, we have, for $k \geq k_0$,

$$\alpha = \frac{p_{k-1}r_k + p_{k-2}}{q_{k-1}r_k + q_{k-2}} = \frac{p_{k+h-1}r_{k+h} + p_{k+h-2}}{q_{k+h-1}r_{k+h} + q_{k+h-2}} = \frac{p_{k+h-1}r_k + p_{k+h-2}}{q_{k+h-1}r_k + q_{k+h-2}},$$

$$(48)$$

so that

$$\frac{p_{k-1}r_k + p_{k-2}}{q_{k-1}r_k + q_{k-2}} = \frac{p_{k+h-1}r_k + p_{k+h-2}}{q_{k+h-1}r_k + q_{k+h-2}}.$$

Thus, the number r_k satisfies a quadratic equation with integral coefficients and, consequently, is a quadratic irrational number. But, in this case, the first inequality of (48) shows that α too is a quadratic irrational number.

The converse is somewhat more complicated. Suppose that the number α satisfies the quadratic equation

$$a\alpha^2 + b\alpha + c = 0 \qquad (49)$$

with integral coefficients. If we write α in terms of its remainders of order n

$$\alpha = \frac{p_{n-1}r_n + p_{n-2}}{q_{n-1}r_n + q_{n-2}}$$

(again using eq. [16] of Chap. I), we see that r_n satisfies the equation

$$A_n r_n^2 + B_n r_n + C_n = 0, \tag{50}$$

where A_n, B_n, and C_n are integers defined by

$$\left.\begin{array}{l} A_n = ap_{n-1}^2 + bp_{n-1}q_{n-1} + cq_{n-1}^2, \\[4pt] B_n = 2ap_{n-1}p_{n-2} + b(p_{n-1}q_{n-2} + p_{n-2}q_{n-1}) + 2cq_{n-1}q_{n-2}, \\[4pt] C_n = ap_{n-2}^2 + bp_{n-2}q_{n-2} + cq_{n-2}^2, \end{array}\right\} \tag{51}$$

from which, in particular, it follows that

$$C_n = A_{n-1}. \tag{52}$$

With these formulas, it is easy to verify directly that

$$B_n^2 - 4A_n C_n = (b^2 - 4ac)(p_{n-1}q_{n-2} - q_{n-1}p_{n-2})^2 = b^2 - 4ac, \tag{53}$$

that is, that the discriminant of (50) is the same for all n and is equal to the discriminant of (49). Furthermore, since

$$\left| \alpha - \frac{p_{n-1}}{q_{n-1}} \right| < \frac{1}{q_{n-1}^2},$$

it follows that

$$p_{n-1} = \alpha q_{n-1} + \frac{\delta_{n-1}}{q_{n-1}} \qquad (|\delta_{n-1}| < 1);$$

therefore, the first formula of (51) gives us

$$A_n = a\left(\alpha q_{n-1} + \frac{\delta_{n-1}}{q_{n-1}}\right)^2 + b\left(\alpha q_{n-1} + \frac{\delta_{n-1}}{q_{n-1}}\right)q_{n-1} + cq_{n-1}^2$$

$$= (a\alpha^2 + b\alpha + c)q_{n-1}^2 + 2a\alpha\delta_{n-1} + a\frac{\delta_{n-1}^2}{q_{n-1}^2} + b\delta_{n-1},$$

from which, on the basis of (49), we have

$$|A_n| = \left| 2a\alpha\delta_{n-1} + a\frac{\delta_{n-1}^2}{q_{n-1}^2} + b\delta_{n-1} \right| < 2|a\alpha| + |a| + |b|.$$

and, on the basis of (52),

$$|C_n| = |A_{n-1}| < 2|a\alpha| + |a| + |b|.$$

Thus, the coefficients A_n and C_n in (50) are bounded in absolute value and hence may assume only a finite number of distinct values as n varies. It then follows on the basis of (53) that B_n may take only a finite number of distinct values. Thus, as n increases from 1 to ∞, we can encounter only a finite number of distinct equations in (50). But, in any case, r_n can take only a finite number of distinct values, and therefore, for properly chosen k and h,

$$r_k = r_{k+h}.$$

This shows that the continued fraction representing a is periodic and thus proves the second assertion of the theorem.

No proofs analogous to this are known for continued fractions representing algebraic irrational numbers of higher degrees. In general, all that is known concerning the approximation of algebraic numbers of higher degrees by rational fractions amounts to some elementary corollaries to Liouville's theorem, and certain newer propositions strengthening it. It is interesting to note that we do not, at the present time, know the continued-fraction expansion of a single algebraic number of degree higher than 2. We do not know, for example, whether the sets of elements in such expansions are bounded or unbounded. In general, questions connected with the continued-fraction expansion of algebraic numbers of higher degree than the second are extremely difficult and have hardly been studied.

Chapter III

THE MEASURE THEORY OF
CONTINUED FRACTIONS

11. Introduction

In the course of the preceding chapter, we saw that real numbers can be quite different in their arithmetic properties. Besides the basic divisions of the real numbers into rational and irrational or algebraic and transcendental numbers, there are several considerably finer subdivisions of these numbers based on a whole series of criteria characterizing their arithmetic nature (most importantly, criteria involving the approximation by rational fractions that these numbers admit). In all these cases, we have, up to now, been content with simple proofs that numbers having certain arithmetic properties actually do exist. Thus, we know that numbers exist admitting approximation by rational fractions of the form p/q with order of accuracy not exceeding $1/q^2$ (for example, all quadratic irrational numbers); but we also know that there exist numbers admitting approximation of much higher order (Theor. 22, Chap. II). The following question naturally arises: which of these two opposite properties should we consider the more "general," that is, which of these two types of real numbers do we "encounter more often"?

If we wish to give a precise formulation of the question just posed, we must remember that each time we refer to some property or other of the real numbers (for example, irrationality, transcendentality, possession of a bounded sequence of elements, etc.), the set of real numbers is partitioned with respect to that property into two sets: (1) the set of numbers possessing that property, and (2) the set of numbers not possessing it. The question is then clearly reduced to a comparative study of these two sets, with the purpose of determining which of them contains more numbers. However, sets of real numbers can be compared with each other from various points of view, and in terms of various characteristics. We can pose the question of their cardinality, of their measure, or of a number of other gauges. As regards both methods and results, the study of the *measure* of sets of numbers defined by a given property of their elements has proven the most interesting. This study, which we shall call the *measure arithmetic of the continuum*, has undergone considerable development in recent years, and has led to a large number of simple and interesting principles. As with every study of the arithmetic nature of irrational numbers, the apparatus of continued fractions is the most natural and the best investigating instrument. However, to make this apparatus an instrument for measure arithmetic (that is, to apply it to the study of the *measure* of sets whose members are defined by some arithmetic property), we must first subject the apparatus itself to a detailed analysis from all aspects. We must, in other words, learn to determine the measure of numbers whose expansions in continued fractions possess some previously stated property. Questions of this kind can be quite varied: we may inquire about the measure of the set of numbers for which $a_4 = 2$, or for which q_{10} is less than 1,000, or which have a bounded sequence of elements, or which have no even elements, and so on. The methods used in solving problems such as these constitute the *measure theory of continued fractions*. It is to the fundamentals and the elementary applications of this theory that the present chapter is devoted.

Since the addition of an integer to a given real number does not change the fundamental properties of that real number, we shall henceforth confine ourselves to an examination of the real numbers between zero and one; that is, we shall always assume that $a_0 = 0$. Such a restriction to a finite interval is necessary in measure theory if we do not wish the measure of a set, in the general case, to be infinite. We are assuming that the reader is familiar with the basic propositions of measure theory.[1]

12. The elements as functions of the number represented

Every real number α has a *unique expansion* as a continued fraction

$$\alpha = [a_0; \ a_1, \ a_2, \ \ldots];$$

each element a_n is therefore uniquely defined by the number α; that is, it is a single-valued function of α:

$$a_n = a_n(\alpha).$$

---

[1] The material contained in any text on functions of a real variable will be more than sufficient for understanding this chapter.

To develop the measure theory of continued fractions, we must first study the properties of this function and obtain a general picture of its behavior. The present section is devoted to this problem.

As we noted in section 11, we are henceforth assuming that $a_0 = 0$. For simplicity in notation, we shall always write

$$\alpha = [a_1, a_2, \ldots]$$

instead of

$$\alpha = [a_0; a_1, a_2, \ldots].$$

Thus,

$$[a_1, a_2, \ldots] = \cfrac{1}{a_1 + \cfrac{1}{a_2 + \ldots}}$$

Let us begin by examining the first element a_1 as a function of α. Since

$$\frac{1}{\alpha} = a_1 + \cfrac{1}{a_2 + \ldots},$$

it is obvious that $a_1 = [1/\alpha]$, that is, a_1 is the greatest integer not exceeding $1/\alpha$. Thus,

$$a_1 = 1, \quad \text{for } 1 \leqslant \frac{1}{\alpha} < 2; \quad \text{i.e., } \frac{1}{2} < \alpha \leqslant 1,$$

$$a_1 = 2, \quad \text{for } 2 \leqslant \frac{1}{\alpha} < 3; \quad \text{i.e., } \frac{1}{3} < \alpha \leqslant \frac{1}{2},$$

$$a_1 = 3, \quad \text{for } 3 \leqslant \frac{1}{\alpha} < 4; \quad \text{i.e., } \frac{1}{4} < \alpha \leqslant \frac{1}{3}, \quad \text{etc.}$$

In general,

$$a_1 = k, \quad \text{for } k \leqslant \frac{1}{\alpha} < k+1; \quad \text{i.e., } \frac{1}{k+1} < \alpha \leqslant \frac{1}{k}.$$

The function $a_1 = a_1(\alpha)$ thus has a discontinuity at all those values of α for which $1/\alpha$ is an integer, and it increases without bound as α approaches 0. Figure 1 gives a graphical representation. We note that a_1 is constant throughout each of the intervals

$$\frac{1}{k+1} < \alpha \leqslant \frac{1}{k}.$$

We shall call these intervals *intervals of the first rank*. We note also that

$$\int_0^1 a_1(\alpha)\,d\alpha = +\infty,$$

since this integral is obviously equivalent to the divergent series

$$\sum_{k=1}^\infty k\left(\frac{1}{k} - \frac{1}{k+1}\right) = \sum_{k=1}^\infty \frac{1}{k+1}.$$

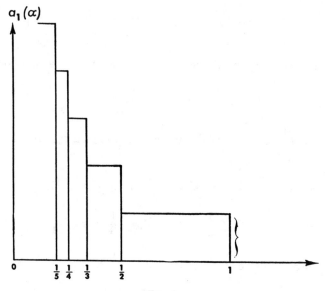

Fig. 1

Let us now examine the function $a_2(\alpha)$, first considering some fixed interval of the first rank:

$$\frac{1}{k+1} < \alpha \leqslant \frac{1}{k}.$$

In this interval, $a_1 = k$ everywhere and, consequently,

$$\alpha = \frac{1}{k + \dfrac{1}{r_2}},$$

where $1 \leq r_2 < \infty$ and $a_2 = [r_2]$. As r_2 increases from 1 to ∞, a increases from $1/(k+1)$ to $1/k$, thus taking all values in the given interval of the first rank. It is then obvious that

$$a_2 = 1, \quad \text{for } 1 \leqslant r_2 < 2; \quad \text{i.e.,} \quad \frac{1}{k+1} \leqslant a < \frac{1}{k+\frac{1}{2}},$$

$$a_2 = 2, \quad \text{for } 2 \leqslant r_2 < 3; \quad \text{i.e.,} \quad \frac{1}{k+\frac{1}{2}} \leqslant a < \frac{1}{k+\frac{1}{3}},$$

$$a_2 = 3, \quad \text{for } 3 \leqslant r_2 < 4; \quad \text{i.e.,} \quad \frac{1}{k+\frac{1}{3}} \leqslant a < \frac{1}{k+\frac{1}{4}},$$

and, in general,

$$a_2 = l, \quad \text{for } l \leqslant r_2 < l+1, \quad \text{i.e.,} \quad \frac{1}{k+\frac{1}{l}} \leqslant a < \frac{1}{k+\frac{1}{l+1}}.$$

Thus, in this fixed first-rank interval, the graphical representation of the function $a_2(a)$ has the form shown in Figure 2.

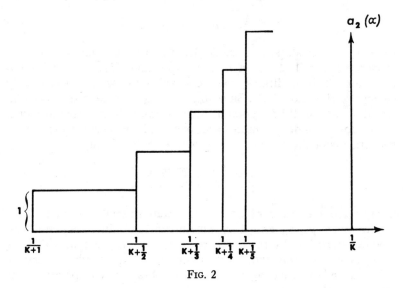

FIG. 2

The function $a_2(\alpha)$ is constant in each of the intervals

$$\left(\frac{1}{k + \dfrac{1}{l}} \,, \; \frac{1}{k + \dfrac{1}{l+1}} \right),$$

which we shall call intervals of the second rank. Every interval of the first rank can consequently be partitioned into a countable set of intervals of the second rank, going from left to right. (We recall that intervals of the first rank form a sequence going from right to left.) The set of points at which $a_1 = k$ is an interval of the first rank. The set of points at which $a_2 = l$ is a countable set of intervals of the second rank (one in each of the intervals of the first rank). Each interval of the first rank is defined by a condition of the form $a_1 = k$, and each interval of the second rank by conditions of the form $a_1 = k$, $a_2 = l$.

Suppose that we have defined all intervals of rank n and that we are investigating the set of functions $a_1(\alpha)$, $a_2(\alpha)$, $\cdots$, $a_n(\alpha)$. Each system of values

$$a_1 = k_1, \quad a_2 = k_2, \quad \ldots, \quad a_n = k_n \tag{54}$$

defines some interval J_n of rank n. To investigate the behavior of the function $a_{n+1}(\alpha)$ in the interval J_n, we note that an arbitrary number α of this interval may be represented in the form

$$\alpha = [k_1, \, k_2, \, \ldots, \, k_n, \, r_{n+1}], \tag{55}$$

where r_{n+1} takes all possible values from 1 to ∞. Conversely, for arbitrary r_{n+1} (where $1 < r_{n+1} < \infty$), the expression (55) gives us the number α for which conditions (54) are satisfied and which consequently belongs to the interval J_n. Since $a_{n+1} = [r_{n+1}]$, we see that in each interval of rank n, the function $a_{n+1}(\alpha)$ assumes all the integral values from 1 to ∞. To draw a more exact picture, let us agree to denote the convergents of the number α, as usual, by p_k/q_k. Then,

$$\alpha = \frac{p_n r_{n+1} + p_{n-1}}{q_n r_{n+1} + q_{n-1}},$$

where r_{n+1} increases from 1 to ∞ as α runs through the interval J_n. The numbers p_n, q_n, p_{n-1}, q_{n-1} remain constant, since they are completely determined by the numbers a_1, a_2, $\cdots$, a_n, each of which has the same value for all points of the interval J_n. In particular, by setting $r_{n+1} = 1$ and then letting $r_{n+1} \to \infty$, we obtain as end points of the interval J_n the points

$$\frac{p_n + p_{n-1}}{q_n + q_{n-1}} \qquad \text{and} \qquad \frac{p_n}{q_n};$$

and since

$$\alpha - \frac{p_n}{q_n} = \frac{p_n r_{n+1} + p_{n-1}}{q_n r_{n+1} + q_{n-1}} - \frac{p_n}{q_n} = \frac{(-1)^n}{q_n (q_n r_{n+1} + q_{n-1})},$$

α is a monotonic function of r_{n+1} in the interval $(1, \infty)$. Conversely, r_{n+1}—and hence, a_{n+1}—is a monotonic function of α in the interval

$$J_n = \left(\frac{p_n}{q_n}, \ \frac{p_n + p_{n-1}}{q_n + q_{n-1}} \right);$$

thus, as α runs through the interval J_n, the function $a_{n+1}(\alpha)$ takes, in succession, the values $1, 2, 3, \cdots$, partitioning the interval J_n into a countable set of intervals of rank $n + 1$. This sequence is taken from right to left for even n and from left to right for odd n.

Thus, the function $a_n(\alpha)$, at least qualitatively, is completely defined. Let us agree to call the interval $(0, 1)$ the (unique) interval of rank zero, and let us cover it with a net of finer intervals, placing in every already constructed interval of rank n a sequence of intervals of rank $n + 1$. This sequence is taken from right to left if n is even and from left to right if n is odd. The function $a_{n+1}(\alpha)$ (for $n = 0, 1, 2, \cdots$) is constant in each of these intervals of rank $n + 1$. This function is monotonic and takes all integral values from 1 to ∞ in each interval of rank n. To each system of values

$$a_1 = k_1, \quad a_2 = k_2, \ \ldots, \quad a_n = k_n$$

there corresponds a uniquely defined interval of rank n, and vice versa. The more general system of values

$$a_{m_1} = k_1, \quad a_{m_2} = k_2, \ \ldots, \quad a_{m_s} = k_s$$

determines, generally speaking, a countable set of intervals.

The first question posed by the measure theory of continued fractions naturally consists in determining the measure of the set of those points of the interval $(0, 1)$ for which $a_n = k$. We already know that this set is a union of disjoint intervals. It is then a question of evaluating the sum of these intervals. A first approximation to the solution of this problem is obtained quite easily.

From here on, let us agree to denote by

$$E \begin{pmatrix} n_1, \ n_2, \ \ldots, \ n_s \\ k_1, \ k_2, \ \ldots, \ k_s \end{pmatrix}$$

the set of points of the interval $(0, 1)$ for which the following conditions are satisfied:

$$a_{n_1} = k_1, \quad a_{n_2} = k_2, \quad \ldots, \quad a_{n_s} = k_s;$$

here, of course, all the n_i and k_i are natural numbers and the n_i are all different from each other. We already know that such a set is always a union of intervals. In particular, the set

$$E \begin{pmatrix} 1, & 2, \ldots, & n \\ k_1, & k_2, \ldots, & k_n \end{pmatrix}$$

is, as we know, an interval of rank n, characterized by the relationships

$$a_i = k_i \qquad (i = 1, 2, \ldots, n).$$

Obviously, we always have

$$\sum_{k_l=1}^{\infty} E \begin{pmatrix} n_1, \ldots, & n_{l-1}, & n_l, & n_{l+1}, \ldots, & n_s \\ k_1, \ldots, & k_{l-1}, & k_l, & k_{l+1}, \ldots, & k_s \end{pmatrix} \tag{56}$$

$$= E \begin{pmatrix} n_1, \ldots, & n_{l-1}, & n_{l+1}, \ldots, & n_s \\ k_1, \ldots, & k_{l-1}, & k_{l+1}, \ldots, & k_s \end{pmatrix}.$$

Finally, let us agree to denote by $\mathfrak{M}E$ the measure of the set E. Let us consider an arbitrary interval

$$J_n = E \begin{pmatrix} 1, & 2, & \ldots, & n \\ k_1, & k_2, & \ldots, & k_n \end{pmatrix}$$

of rank n containing the interval

$$J_{n+1}^{(s)} = E \begin{pmatrix} 1, & 2, & \ldots, & n, & n+1 \\ k_1, & k_2, & \ldots, & k_n, & s \end{pmatrix}$$

of rank $n + 1$. We already know that the end points of the interval J_n are the points

$$\frac{p_n}{q_n}$$

and

$$\frac{p_n + p_{n-1}}{q_n + q_{n-1}},$$

where p_k/q_k denotes the convergent of order k of the continued fraction

$$[k_1, k_2, \ldots, k_n].$$

On the other hand, for all points of the interval $J_{n+1}^{(s)}$, we have

$$a_{n+1} = [r_{n+1}] = s,$$

and hence,

$$s \leqslant r_{n+1} < s+1.$$

Thus, among all the points

$$\alpha = \frac{p_n r_{n+1} + p_{n-1}}{q_n r_{n+1} + q_{n-1}}$$

of the interval J_n, those for which $s \leq r_{n+1} < s+1$ belong to the interval $J_{n+1}^{(s)}$. Hence, it follows, in particular, that the end points of the interval $J_{n+1}^{(s)}$ will be the points

$$\frac{p_n s + p_{n-1}}{q_n s + q_{n-1}} \quad \text{and} \quad \frac{p_n(s+1)+p_{n-1}}{q_n(s+1)+q_{n-1}}.$$

Therefore,

$$\mathfrak{M}J_n = \left| \frac{p_n}{q_n} - \frac{p_n+p_{n-1}}{q_n+q_{n-1}} \right| = \frac{1}{q_n(q_n+q_{n-1})} = \frac{1}{q_n^2\left(1+\frac{q_{n-1}}{q_n}\right)},$$

$$\mathfrak{M}J_{n+1}^{(s)} = \left| \frac{p_n s + p_{n-1}}{p_n s + q_{n-1}} - \frac{p_n(s+1)+p_{n-1}}{q_n(s+1)+q_{n-1}} \right|$$

$$= \frac{1}{[q_n s + q_{n-1}][q_n(s+1)+q_{n-1}]}$$

$$= \left| \frac{1}{q_n^2 s^2\left(1+\frac{q_{n-1}}{sq_n}\right)\left(1+\frac{1}{s}+\frac{q_{n-1}}{sq_n}\right)} \right|,$$

and, consequently,

$$\frac{\mathfrak{M}J_{n+1}^{(s)}}{\mathfrak{M}J_n} = \frac{1}{s^2} \frac{1+\frac{q_{n-1}}{q_n}}{\left(1+\frac{q_{n-1}}{sq_n}\right)\left(1+\frac{1}{s}+\frac{q_{n-1}}{sq_n}\right)}.$$

Here, the second factor on the right side is obviously always less than 2 and greater than $\frac{1}{3}$. (The last assertion results from the facts that

$$\frac{1+\frac{q_{n-1}}{q_n}}{1+\frac{q_{n-1}}{sq_n}} \geqslant 1 \quad \text{and} \quad 1+\frac{1}{s}+\frac{q_{n-1}}{sq_n} < 3.)$$

Therefore, we obtain

$$\frac{1}{3s^2} < \frac{\mathfrak{M}J_{n+1}^{(s)}}{\mathfrak{M}J_n} < \frac{2}{s^2}. \tag{57}$$

This shows that, given an arbitrary interval of rank n, the interval of rank $n + 1$ characterized by the value $a_{n+1} = s$ occupies a part of the given interval of the order of $1/s^2$. The fact that the bounds given by the inequalities in (57) are completely independent not only of the numbers $k_1, k_2, \cdots, k_n$ but also of the rank n (and are determined exclusively by the number s) is extremely important. If we rewrite these inequalities in the form

$$\frac{\mathfrak{M}J_n}{3s^2} < \mathfrak{M}J_{n+1}^{(s)} < \frac{2\mathfrak{M}J_n}{s^2},$$

sum over all intervals J_n of rank n (or, equivalently, over $k_1, k_2, \cdots,$ k_n from 1 to ∞), and, finally, note that

$$\sum \mathfrak{M}J_n = 1, \qquad \sum \mathfrak{M}J_{n+1}^{(s)} = \mathfrak{M}E\binom{n+1}{s},$$

we obtain

$$\frac{1}{3s^2} < \mathfrak{M}E\binom{n+1}{s} < \frac{2}{s^2}.$$

This provides a first approximation to the solution to the problem being considered. We see that the measure of the set of points at which some definite element has a given value s always lies between $1/3s^2$ and $2/s^2$ (and consequently, is a quantity of the order $1/s^2$).

13. Measure-theoretic evaluation of the increase in the elements

We now have the necessary tools for solving problems involving the measure of sets containing an infinite number of elements. As a first example of such a problem, we shall prove the following simple theorem.

THEOREM 29. *The set of all numbers in the interval $(0, 1)$ with bounded elements is of measure zero.*

PROOF. We denote by E_M the set of numbers in the interval $(0, 1)$

all of whose elements are less than M. Let J_n be any interval of rank n whose points satisfy the conditions

$$a_i < M \qquad (i = 1, 2, \ldots, n). \tag{58}$$

The points of the interval J_n that satisfy the additional condition $a_{n+1} = k$ form an interval of rank $n + 1$. We denote this interval by $J_{n+1}^{(k)}$. From the first of the inequalities in (57),

$$\mathfrak{M} J_{n+1}^{(k)} > \frac{1}{3k^2} \, \mathfrak{M} J_n,$$

so that

$$\mathfrak{M} \sum_{k \geqslant M} J_{n+1}^{(k)} > \frac{1}{3} \mathfrak{M} J_n \sum_{k \geqslant M} \frac{1}{k^2}$$

$$> \frac{1}{3} \mathfrak{M} J_n \sum_{i=1}^{\infty} \frac{1}{(M+i)^2} > \frac{1}{3} \mathfrak{M} J_n \int_{M+1}^{\infty} \frac{du}{u^2} = \frac{1}{3(M+1)} \mathfrak{M} J_n,$$

and since

$$\sum_{k=1}^{\infty} J_{n+1}^{(k)} = J_n,$$

it follows that

$$\mathfrak{M} \sum_{k < M} J_{n+1}^{(k)} > \left\{ 1 - \frac{1}{3(M+1)} \right\} \mathfrak{M} J_n = \tau \mathfrak{M} J_n, \tag{59}$$

where

$$\tau = 1 - \frac{1}{3(M+1)};$$

here, obviously, $\tau < 1$ if $M > 0$.

If we denote by $E_M^{(n)}$ the set of numbers of the interval $(0, 1)$ characterized by conditions (58), we see from inequality (59) that the measure of that portion of the set $E_M^{(n+1)}$ contained in some interval J_n of rank n is less than that of $\tau \mathfrak{M} J_n$. Since, obviously, an interval of rank n that does not belong to the set $E_M^{(n)}$ (that is, one that does not satisfy the conditions in eq. [58]) cannot contain any point of the set $E_M^{(n+1)}$, if we sum inequality (59) over all intervals of rank n in the set $E_M^{(n)}$, we obtain

$$\mathfrak{M} E_M^{(n+1)} < \tau \mathfrak{M} E_M^{(n)}. \tag{60}$$

Successive application of this inequality gives us

$$\mathfrak{M}E_M^{(n+1)} < \tau^n \mathfrak{M}E_M^{(1)} \qquad (n \geqslant 1).$$

so that

$$\mathfrak{M}E_M^{(n)} \to 0 \qquad \text{as} \qquad n \to \infty,$$

since $\tau < 1$. But the set E_M that we have defined above is obviously contained in each of the sets $E_M^{(n)}$. Consequently,

$$\mathfrak{M}E_M = 0.$$

Now setting

$$\sum_{M=1}^{\infty} E_M = E,$$

we obtain

$$\mathfrak{M}E \leqslant \sum_{M=1}^{\infty} \mathfrak{M}E_M = 0.$$

But every number with bounded elements obviously belongs to the set E_M for sufficiently large M and, hence, belongs to the set E, which proves the theorem.

We know (Theor. 23, Chap. II) that numbers with bounded elements are those numbers α that do not admit an approximation by rational fractions better than in accordance with the law

$$\left| \alpha - \frac{p}{q} \right| < \frac{C}{q^2}. \tag{61}$$

(We note that among these numbers are all quadratic irrationals.) We see now that all these numbers form a set of measure zero. In other words, *almost all* numbers (that is, all but a set of measure zero) admit a best approximation by rational fractions. Evidently, then, the basic problem of the measure theory of approximation is the question of determining the measure of the set of numbers admitting some specified degree of approximation by rational fractions. In particular, what is the best law of approximation admitted by *almost all* (see above) numbers? In other words, within what limits can the law given by inequality (61) be improved if we agree to neglect the set of numbers α which is of measure zero? We shall solve this problem in the next section.

THEOREM 30. *Suppose that $\varphi(n)$ is an arbitrary positive function with natural argument n. The inequality*

$$a_n = a_n(\alpha) \geqslant \varphi(n) \qquad (62)$$

is, for almost all α, satisfied by an infinite number of values of n if the series $\Sigma_{n=1}^{\infty} 1/\varphi(n)$ diverges. On the other hand, this inequality is, for almost all α, satisfied by only a finite number of values of n if the series $\Sigma_{n=1}^{\infty} 1/\varphi(n)$ converges.

PRELIMINARY REMARK. In particular, if we set the function $\varphi(n)$ equal to a constant positive number M, we conclude from Theorem 30 that the set E_M, which we used in proving Theorem 29, is of measure zero. Thus, Theorem 29 can be regarded as one of the simplest cases of Theorem 30.

PROOF. The first assertion of the theorem is proved in a manner completely analogous to that used in the proof of Theorem 29. Suppose that J_{m+n} is an interval of rank $m + n$ at all of whose points

$$a_{m+i} < \varphi(m+i) \qquad (i = 1, 2, \ldots, n). \qquad (63)$$

(We shall not impose any conditions on $a_1, a_2, \cdots, a_m$.) Keeping the notation that we used in the proof of Theorem 29, we obtain the inequality, analogous to inequality (59),

$$\mathfrak{M} \sum_{k < \varphi(m+n+1)} J^{(k)}_{m+n+1} < \left\{ 1 - \frac{1}{3(1 + \varphi(m+n+1))} \right\} \mathfrak{M} J_{m+n}.$$

When we sum this inequality over all intervals of rank $m + n$ that satisfy conditions (63) (denoting the set of all numbers of the interval $(0, 1)$ satisfying these conditions by $E_{m,n}$), we obtain

$$\mathfrak{M} E_{m, n+1} < \left\{ 1 - \frac{1}{3(1 + \varphi(m+n+1))} \right\} \mathfrak{M} E_{m, n}.$$

Successive application of this inequality gives

$$\mathfrak{M} E_{m, n} < \mathfrak{M} E_{m, 1} \prod_{i=2}^{n} \left\{ 1 - \frac{1}{3(1 + \varphi(m+i))} \right\}.$$

If the series $\Sigma_{n=1}^{\infty} 1/\varphi(n)$ diverges, the series

$$\sum_{i=2}^{\infty} \frac{1}{3(1 + \varphi(m+i))}$$

will obviously, for arbitrary m, also diverge. From the theory of infinite products, it follows that the product

$$\prod_{i=2}^{n}\left\{1-\frac{1}{3\left(1+\varphi\left(m+i\right)\right)}\right\}$$

approaches zero as $n \to \infty$. Thus, we have, for arbitrary m,

$$\mathfrak{M}E_{m,\,n} \to 0 \quad \text{as} \quad n \to \infty.$$

But every number a for which

$$a_{m+i} < \varphi(m+i) \qquad (i=1,\ 2,\ \ldots),$$

obviously belongs to all the sets

$$E_{m,\,n} \qquad (n=1,\ 2,\ \ldots);$$

therefore, the set of all these numbers, which we shall denote by E_m, must be of measure zero. Finally, if we set

$$E_1 + E_2 + \cdots + E_m + \cdots = E,$$

we see that $\mathfrak{M}E = 0$. But every number a for which inequality (62) is satisfied only a finite number of times must, obviously, for sufficiently large m, belong to the set E_m and, hence, to the set E. This proves the first assertion of the theorem.

Suppose now that the series $\Sigma_{n=1}^{\infty}\,1/\varphi(n)$ converges. Suppose that J_n is one of the intervals of rank n and that $J_{n+1}^{(k)}$ is an interval of rank $n+1$ contained in J_n and defined by the additional condition that $a_{n+1} = k$. From the second inequality of (57), we have

$$\mathfrak{M}J_{n+1}^{(k)} < \frac{2}{k^2}\,\mathfrak{M}J_n,$$

so that

$$\mathfrak{M}\sum_{k>\varphi(n+1)}J_{n+1}^{(k)} < 2J\mathfrak{M}_n\sum_{k>\varphi(n+1)}\frac{1}{k^2}$$

$$\leqslant 2\mathfrak{M}J_n\sum_{i=0}^{\infty}\frac{1}{\{\varphi(n+1)+i\}^2}$$

$$< 2\mathfrak{M}J_n\left\{\frac{1}{\varphi(n+1)}+\int_{\varphi(n+1)}^{\infty}\frac{du}{u^2}\right\}=\frac{4\mathfrak{M}J_n}{\varphi(n+1)}.$$

If we denote by F_n the set of numbers of the interval $(0, 1)$ for which $a_n \geq \varphi(n)$ and sum the inequality obtained over all intervals J_n of rank n, we see that

$$\mathfrak{M}F_{n+1} < \frac{4}{\varphi(n+1)},$$

since $\Sigma \mathfrak{M}J_n = 1$. Thus, the measures of the sets $F_1, F_2, \cdots, F_n, \cdots$ form a convergent series. Denoting by F the set of those numbers in the interval $(0, 1)$ that belong to an infinite number of sets F_n, we then have[2]

$$\mathfrak{M}F = 0.$$

But the set F is, of course, just the set of numbers for which the inequality (62) is satisfied for an infinite number of values of n. This proves the second assertion of the theorem.

14. Measure-theoretic evaluation of the increase in the denominators of the convergents. The fundamental theorem of the measure theory of approximation

THEOREM 31. *There exists an absolutely positive constant B such that almost everywhere, for sufficiently large n,*

$$q_n = q_n(\alpha) < e^{Bn}.$$

PRELIMINARY REMARK. We saw in section 4 of Chapter I (Theor. 12) that the denominators q_n for *all* numbers α increase with increasing n no more slowly than some geometric progression with absolutely constant ratio. Theorem 31 asserts that, for *almost all* α, the numbers q_n do not increase faster than some other geometric progression, also with absolutely constant ratio. This situation can be expressed in a different way: there exist two absolute constants a and A (where $1 < a < A$) such that, for almost all numbers α in the interval $(0, 1)$, for sufficiently large n,

$$a < \sqrt[n]{q_n} < A.$$

---

[2] This is a well-known theorem in measure theory. However, here is the proof: Obviously, the set F is, for arbitrary m, contained in the set $\Sigma_{n=m}^{\infty}F_n$; the measure of the latter set does not exceed $\Sigma_{n=m}^{\infty}\mathfrak{M}F_n$ and, consequently, for sufficiently large m, it may be made arbitrarily small.

In fact, there is a considerably stronger proposition: there exists an absolute constant γ such that almost everywhere

$$\sqrt[n]{q_n} \to \gamma \qquad (n \to \infty);$$

however, the proof of this theorem is considerably more complicated and requires certain more powerful tools, which we shall discuss in sections 15 and 16. Unfortunately, the framework of the present book does not allow inclusion of this proof.[3] On the other hand, for our immediate purpose, which is to establish Theorem 32, the property of the numbers q_n referred to in Theorem 31 is quite sufficient.

PROOF. We denote by $E_n(g)$ ($n > 0$, $g \geq 1$) the set of numbers in the interval $(0, 1)$ for which

$$a_1 a_2 \ldots a_n \geq g.$$

Obviously, this set represents a system of intervals of rank n. The length of any one of these intervals is, as we know from section 12, equal to

$$\left| \frac{p_n}{q_n} - \frac{p_n + p_{n-1}}{q_n + q_{n-1}} \right| = \frac{1}{q_n(q_n + q_{n-1})} < \frac{1}{q_n^2} < \frac{1}{(a_n a_{n-1} \cdots a_2 a_1)^2},$$

since successive application of the obvious inequality

$$q_n > a_n q_{n-1}$$

gives

$$q_n > a_n a_{n-1} \cdots a_2 a_1.$$

Therefore,

$$\mathfrak{M} E_n(g) < \sum_{a_1 a_2 \ldots a_n \geq g} \frac{1}{a_n^2 a_{n-1}^2 \cdots a_2^2 a_1^2}, \qquad (64)$$

where the summation is taken over all combinations of natural numbers $a_1, a_2, \cdots, a_n$ that satisfy the inequality $a_1 a_2 \cdots a_n \geq g$. To evaluate this sum, we note that

$$\prod_{i=1}^{n} \frac{1}{a_i^2} = \prod_{i=1}^{n} \left(1 + \frac{1}{a_i}\right) \frac{1}{a_i(a_i+1)} \leq 2^n \prod_{i=1}^{n} \frac{1}{a_i(a_i+1)}$$

$$= 2^n \prod_{i=1}^{n} \int_{a_i}^{a_i+1} \frac{dx_i}{x_i^2} = 2^n \int_{a_1}^{a_1+1} \int_{a_2}^{a_2+1} \cdots \int_{a_n}^{a_n+1} \frac{dx_1\, dx_2 \ldots dx_n}{x_1^2 x_2^2 \ldots x_n^2},$$

---

[3] Proof of the above statement was obtained by Khinchin in 1935. See "Zur metrischen Kettenbruchtheorie," *Compositio Mathematica*, **3**, No. 2, 275–285 (1936). Soon afterward, the French mathematician P. Lévy found an explicit expression for the constant γ, namely, $\ln \gamma = \pi^2/(12 \ln 2)$ (see P. Lévy, *Théorie de l'addition des variables aléatoires*, Paris, 1937, p. 320). (B. G.)

and, consequently,

$$\sum_{a_1 a_2 \ldots a_n \geqslant g} \left\{ \prod_{i=1}^{n} \frac{1}{a_i^2} \right\} \leqslant 2^n J_n (g),$$

where $J_n(g)$ is the nth-order integral

$$\int \int \cdots \int \frac{dx_1 \, dx_2 \ldots dx_n}{x_1^2 x_2^2 \ldots x_n^2}$$

over the region

$$x_i \geqslant 1 \qquad (i = 1, 2, \ldots, n),$$

$$x_1 x_2 \ldots x_n \geqslant g.$$

For $g \leq 1$, this region is obviously the region $1 \leq x_i < \infty$ (for $i = 1, 2, \cdots, n$), and we obtain

$$J_n (g) = \left\{ \int_1^\infty \frac{dx}{x^2} \right\}^n = 1 \qquad (g \leqslant 1). \tag{65}$$

Let us now show that, for $g > 1$,

$$J_n (g) = \frac{1}{g} \sum_{i=0}^{n-1} \frac{(\ln g)^i}{i!} \tag{66}$$

For $n = 1$, this equation is of the form

$$\int_0^\infty \frac{dx}{x^2} = \frac{1}{g}$$

and hence is true. Assuming it is true for $n = k$, we have

$$J_{k+1} (g) = \int_1^\infty \frac{dx_{k+1}}{x_{k+1}^2} J_k \left(\frac{g}{x_{k+1}} \right) = \frac{1}{g} \int_0^g J_k (u) \, du$$

$$= \frac{1}{g} \left\{ \int_0^1 J_k (u) \, du + \int_1^g J_k (u) \, du \right\}.$$

If, in the first integral, we substitute the value of $J_k(u)$ given by formula (65) and, in the second, that given by formula (66) (which we are assuming established for $n = k$, $g \geq 1$), we obtain

$$J_{k+1} (g) = \frac{1}{g} \left\{ 1 + \sum_{i=0}^{k-1} \frac{(\ln g)^{i+1}}{(i+1)} \right\} = \frac{1}{g} \sum_{i=0}^{k} \frac{(\ln g)^i}{i!},$$

which proves the assertion. Thus,

$$\mathfrak{M}E_n(g) < \frac{2^n}{g} \sum_{i=0}^{n-1} \frac{(\ln g)^i}{i!}.$$

In particular, if we set $g = e^{An}$, where $A > 1$ is a constant, we have

$$\mathfrak{M}E_n(e^{An}) < e^{n(\ln 2 - A)} \sum_{l=0}^{n-1} \frac{(An)^l}{l!}.$$

It is easy to see that in this sum each term is less than

$$\frac{(An)^n}{n!};$$

therefore, if we use Stirling's formula for approximating the factorial, we obtain

$$\mathfrak{M}E_n(e^{An}) < e^{n(\ln 2 - A)} n \frac{(An)^n}{n!}$$

$$< C_1 e^{n(\ln 2 - A)} \frac{n(An)^n}{n^n e^{-n} \sqrt{n}} < C_2 \sqrt{n} e^{-n(A - \ln A - \ln 2 - 1)},$$

where C_1 and C_2 are absolute constants.

But if A is sufficiently large,

$$A - \ln A - \ln 2 - 1 > 0,$$

and, consequently $\mathfrak{M}E_n(e^{An})$ is less than the nth term of some convergent series. Since the series

$$\sum_{n=1}^{\infty} \mathfrak{M}E_n(e^{An})$$

converges, every number in the interval $(0, 1)$, with the exception of a set of measure zero, belongs to only a finite number of the sets $E_n(e^{An})$. This means that for almost all numbers in the interval $(0, 1)$, we must have, for sufficiently large n,

$$a_1 a_2 \ldots a_n < e^{An};$$

also, since

$$q_n = a_n q_{n-1} + q_{n-2} < 2 a_n q_{n-1}$$

and, consequently,

$$q_n < 2^n a_n a_{n-1} \ldots a_2 a_1.$$

it follows that almost everywhere, for sufficiently large n,

$$q_n < 2^n e^{An} = e^{Bn},$$

where $B = A + \ln 2$. This completes the proof of Theorem 31.

This result, which in itself is of considerable interest, is especially important for us at the moment, since we can use it to obtain a simple solution to the basic problem of the measure theory of approximation.

THEOREM 32. *Suppose that $f(x)$ is a positive continuous function of a positive variable x and that $xf(x)$ is a non-increasing function. Then, the inequality*

$$\left| \alpha - \frac{p}{q} \right| < \frac{f(q)}{q} \tag{67}$$

has, for almost all α, an infinite number of solutions in integers p and q (with $q > 0$) if, for some positive c, the integral

$$\int_c^\infty f(x)\,dx \tag{68}$$

diverges. On the other hand, inequality (67) has, for almost all α, only a finite number of solutions in integers p and q (with $q > 0$) if the integral (68) converges.

PRELIMINARY REMARK. In particular, on the basis of Theorem 32, the inequality

$$\left| \alpha - \frac{p}{q} \right| < \frac{1}{q^2 \ln q}$$

has, almost everywhere, an infinite number of solutions. On the other hand, the inequality

$$\left| \alpha - \frac{p}{q} \right| < \frac{1}{q^2 \ln^{1+\epsilon} q}$$

has, for every constant $\epsilon > 0$, almost everywhere, only a finite number of solutions. From these facts, we can get an approximate idea of what changes to expect in the general law of approximation if we agree to neglect a set of measure zero.

PROOF. *Part 1.* Suppose that integral (68) diverges. Let us define

$$\varphi(x) = e^{Bx} f(e^{Bx}),$$

where B is the constant referred to in Theorem 31. Then, the integral

$$\int_a^A \varphi(x)\,dx = \frac{1}{B} \int_{Ba}^{BA} f(u)\,du,$$

where $A > a > 0$, increases without bound as $A \to \infty$. Since the function $\varphi(x)$ is, by hypothesis, non-increasing, the series

$$\sum_{n=1}^{\infty} \varphi(n)$$

diverges. On the basis of Theorem 30, we now conclude that, almost everywhere, the inequality

$$a_{l+1} \geqslant \frac{1}{\varphi(l)}$$

is satisfied for an infinite set of values of i. But when this inequality is satisfied,

$$\left| \alpha - \frac{p_i}{q_i} \right| \leqslant \frac{1}{q_i q_{i+1}} \leqslant \frac{1}{a_{i+1} q_i^2} \leqslant \frac{\varphi(l)}{q_i^2}. \qquad (69)$$

On the basis of Theorem 31, we have, almost everywhere, for sufficiently large i,

$$q_i < e^{Bi},$$

so that

$$i > \frac{\ln q_l}{B}.$$

Therefore, inequality (69) almost everywhere implies the inequality

$$\left| \alpha - \frac{p_i}{q_i} \right| \leqslant \frac{\varphi\left(\dfrac{\ln q_i}{B}\right)}{q_i^2} = \frac{f(q_i)}{q_i},$$

for sufficiently large i. This inequality is satisfied almost everywhere for an infinite set of values of i. This proves the first assertion.

Part 2. Let us now suppose that integral (68), and hence the series,

$$\sum_{n=1}^{\infty} f(n),$$

converges. Let us denote by E_n the set of numbers α in the interval $(0, 1)$ that, for a suitably chosen integer k, satisfy the inequality

$$\left| \alpha - \frac{k}{n} \right| < \frac{f(n)}{n}.$$

(Obviously, the set E_n consists of the set of intervals of length $2f[n]/n$, with centers at the points $1/n, 2/n, \cdots, [n-1]/n$ and of the intervals $\{0, f[n]/n\}$ and $\{1 - f[n]/n, 1\}$.) We then have

$$\mathfrak{M}E_n \leqslant 2f(n).$$

(The symbol $<$ holds if $f[n] > \frac{1}{2}$.) Thus, the series

$$\sum_{n=1}^{\infty} \mathfrak{M}E_n$$

converges. We conclude from this, just as we have done on previous occasions, that almost every number α in the interval $(0, 1)$ can belong to only a finite number of sets E_n. This means that almost all the numbers α in the interval $(0, 1)$ satisfy the inequality

$$\left| \alpha - \frac{p}{q} \right| \geqslant \frac{f(q)}{q}$$

for a sufficiently large positive integer q and for an arbitrary integer p. This proves the second assertion of the theorem.

In the next section, we shall learn a method that makes it possible to solve much more profound problems in the measure theory of continued fractions.

15. Gauss's problem and Kuz'min's theorem

The problem that we are about to discuss was, historically, the first problem in the measure theory of continued fractions. This problem, posed by Gauss, was not solved until 1928.[4]

Setting, as usual

$$\alpha = [0; \, a_1, \, a_2, \, \ldots, \, a_n, \, \ldots],$$
$$r_n = r_n(\alpha) = [a_n; \, a_{n+1}, \, a_{n+2}, \, \ldots],$$

---

[4] See R. O. Kuz'min, "Ob odnoĭ zadache Gaussa" (a problem of Gauss), *Doklady akad. nauk, ser. A*, 375–380 (1928). Another solution was published in the article by P. Lévy, "Sur les lois de probabilité dont dependent les quotients complets et incomplets d'une fraction continue," *Bull. Soc. Math.*, **57**, 178–194 (1929). (B. G.)

we denote by $z_n = z_n(a)$ the value of the continued fraction

$$[0; \; a_{n+1}, \; a_{n+2}, \; \ldots],$$

that is, we set

$$z_n = r_n - a_n.$$

Obviously, we always have

$$0 \leqslant z_n < 1.$$

We denote by $m_n(x)$ the measure of the set of numbers a in the interval $(0, 1)$ for which

$$z_n(a) < x.$$

In one of his letters to Laplace, Gauss stated that he had succeeded in proving a theorem that implied that

$$\lim_{n \to \infty} m_n(x) = \frac{\ln(1+x)}{\ln 2} \qquad (0 \leqslant x \leqslant 1).$$

In his letter, he indicated that it would be very desirable to obtain an estimate for the difference

$$m_n(x) - \frac{\ln(1+x)}{\ln 2} \tag{70}$$

for large values of n, but that he had been unable to do so. Apparently, Gauss' proof was never published, nor were other proofs of his assertion known before 1928, when Kuz'min published his proof and gave a good estimate for the difference (70). The present section is devoted to an exposition of these results and of certain generalizations of them that we shall need later.[5]

It was already known to Gauss that the sequence of functions

$$m_0(x), \; m_1(x), \; m_2(x), \; \ldots, \; m_n(x), \; \ldots$$

satisfies the functional equation

$$m_{n+1}(x) = \sum_{k=1}^{\infty} \left\{ m_n\left(\frac{1}{k}\right) - m_n\left(\frac{1}{k+x}\right) \right\} \; (0 \leqslant x \leqslant 1, \; n \geqslant 0). \tag{71}$$

---

[5] Like Gauss, Kuz'min formulated the results in probability-theory terms, which, of course, does not affect their content from the standpoint of measure.

To show this, note that, on the basis of the obvious relationship

$$z_n = \frac{1}{a_{n+1} + z_{n+1}},$$

the inequality

$$z_{n+1} < x$$

is satisfied if, and only if, for a suitably chosen positive integer k,

$$\frac{1}{k+x} < z_n \leqslant \frac{1}{k}.$$

Since the measure of the set of numbers satisfying this inequality is obviously

$$m_n\left(\frac{1}{k}\right) - m_n\left(\frac{1}{k+x}\right),$$

relationship (71) holds.

It can easily be verified directly that the function

$$\varphi(x) = C \ln(1 + x)$$

satisfies the equation

$$\varphi(x) = \sum_{k=1}^{\infty} \left\{ \varphi\left(\frac{1}{k}\right) - \varphi\left(\frac{1}{k+x}\right) \right\}$$

for an arbitrary constant C, which probably helped Gauss in finding the proper expression for the limit of the function $m_n(x)$ as $n \to \infty$.

Formal differentiation of equation (71) gives

$$m'_{n+1}(x) = \sum_{k=1}^{\infty} \frac{1}{(k+x)^2} m'_n\left(\frac{1}{k+x}\right). \tag{72}$$

The validity of (72) can easily be shown in a rigorous manner. Since obviously $z_0(a) = a$, we have $m_0(x) = x$, and hence $m'_0(x) = 1$. If the function $m'_n(x)$ is, in general, bounded and continuous for some n, the series on the right side of (72) converges uniformly in the interval $(0, 1)$. The sum of this series is therefore bounded, continuous, and equal to $m'_{n+1}(x)$ (due to the well-known theorem on termwise differentiation of series). Thus, (72) is proved inductively.

Equation (72) is much more convenient for making investigations than is equation (71). Kuz'min's basic result, which we shall now prove, has to do with this relationship.

THEOREM 33. *Suppose that* $f_1(x), f_2(x), \cdots, f_n(x), \cdots$ *is a sequence of real functions defined on the interval* $(0, 1)$ *that, on that interval, satisfy the relationship*

$$f_{n+1}(x) = \sum_{k=1}^{\infty} \frac{1}{(k+x)^2} f_n\left(\frac{1}{k+x}\right) \qquad (n \geqslant 0). \qquad (73)$$

If, for $0 \leq x \leq 1$,

$$0 < f_0(x) < M$$

and

$$|f_0'(x)| < \mu,$$

then,

$$f_n(x) = \frac{a}{1+x} + \theta A e^{-\lambda\sqrt{n}} \qquad (0 \leqslant x \leqslant 1),$$

where

$$a = \frac{1}{\ln 2} \int_0^1 f_0(z)\, dz, \qquad |\theta| < 1,$$

λ *is an absolute positive constant, and* A *is a positive constant depending only on* M *and* μ.

The proof is complicated, and therefore we shall first give several elementary lemmas.

LEMMA 1. *For arbitrary* $n \geq 0$,

$$f_n(x) = \sum^{(n)} f_0\left(\frac{p_n + x p_{n-1}}{q_n + x q_{n-1}}\right) \frac{1}{(q_n + x q_{n-1})^2}, \qquad (74)$$

where $(p_n/q, (p_n + p_{n-1})/(q_n + q_{n-1}))$ *is an arbitrary interval of rank* n *and the summation takes place over all intervals of rank* n *(or, what is the same thing, over the elements* $a_1, a_2, \cdots, a_n$ *from* 1 *to* ∞ *).*

PROOF. For $n = 0$, (74) is trivial, because in that case there is a unique interval $(0, 1)$ for which $p_0 = 0$, $q_0 = 1$, $p_{-1} = 1$, and $q_{-1} = 0$. Assuming now that equation (74) is valid for some n, we have, on the basis of the fundamental equation (73),

$$f_{n+1}(x) = \sum_{k=1}^{\infty} \frac{1}{(k+x)^2} f_n\left(\frac{1}{k+x}\right)$$

$$= \sum_{k=1}^{\infty} \frac{1}{(k+x)^2} \sum^{(n)} f_0 \left(\frac{p_n + \frac{1}{k+x} p_{n-1}}{q_n + \frac{1}{k+x} q_{n-1}} \right) \frac{1}{\left(q_n + \frac{1}{k+x} q_{n-1} \right)^2}$$

$$= \sum^{(n)} \sum_{k=1}^{\infty} f_0 \left\{ \frac{(p_n k + p_{n-1}) + x p_n}{(q_n k + q_{n-1}) + x q_n} \right\} \frac{1}{\{(q_n k + q_{n-1}) + x q_n\}^2}$$

$$= \sum^{(n+1)} f_0 \left(\frac{p_{n+1} + x p_n}{q_{n+1} + x q_n} \right) \frac{1}{(q_{n+1} + x q_n)^2},$$

which proves the lemma.

LEMMA 2. *Under the conditions of Theorem 33, for $n \geq 0$,*

$$|f'_n(x)| < \frac{\mu}{2^{n-3}} + 4M.$$

PROOF. If we differentiate (74) termwise, we obtain

$$f'_n(x) = \sum^{(n)} f'_0(u) \frac{(-1)^{n-1}}{(q_n + x q_{n-1})^4} - 2 \sum^{(n)} f_0(u) \frac{q_{n-1}}{(q_n + x q_{n-1})^3},$$

where

$$u = \frac{p_n + x p_{n-1}}{q_n + x q_{n-1}}.$$

The validity of termwise differentiation follows from the uniform convergence of both sums on the right side for $0 \leq x \leq 1$. We note that

$$\frac{1}{(q_n + x q_{n-1})^2} < \frac{2}{q_n (q_n + q_{n-1})}.$$

Then, on the basis of Theorem 12 of Chapter I,

$$q_n (q_n + q_{n-1}) > q_n^2 > 2^{n-1},$$

and, in view of the obvious relationship

$$\sum^{(n)} \frac{1}{q (q_n + q_{n-1})} = \sum^{(n)} \left| \frac{p_n}{q_n} - \frac{p_n + p_{n-1}}{q_n + q_{n-1}} \right| = 1,$$

we have, because of the conditions of Theorem 33,

$$|f'_n(x)| < \frac{\mu}{2^{n-3}} + 4M,$$

which was to be proved.

LEMMA 3. *If*

$$\frac{t}{1+x} < f_n(x) < \frac{T}{1+x} \qquad (0 \leqslant x \leqslant 1),$$

it follows that

$$\frac{t}{1+x} < f_{n+1}(x) < \frac{T}{1+x} \qquad (0 \leqslant x \leqslant 1).$$

PROOF. Under the conditions of the lemma, the fundamental equation (73) gives

$$\sum_{k=1}^{\infty} \frac{t}{1+\frac{1}{k+x}} \frac{1}{(k+x)^2} < f_{n+1}(x) < \sum_{k=1}^{\infty} \frac{T}{1+\frac{1}{k+x}} \frac{1}{(k+x)^2},$$

or

$$t \sum_{k=1}^{\infty} \frac{1}{(k+x)(k+x+1)} < f_{n+1}(x) < T \sum_{k=1}^{\infty} \frac{1}{(k+x)(k+x+1)},$$

or, equivalently,

$$t \sum_{k=1}^{\infty} \left(\frac{1}{k+x} - \frac{1}{k+x+1} \right) < f_{n+1}(x) < T \sum_{k=1}^{\infty} \left(\frac{1}{k+x} - \frac{1}{k+x+1} \right),$$

or, finally,

$$\frac{t}{1+x} < f_{n+1}(x) \leqslant \frac{T}{1+x},$$

which was to be proved.

LEMMA 4.

$$\int_0^1 f_n(z)\, dz = \int_0^1 f_0(z)\, dz \qquad (n = 0,\ 1,\ 2,\ \ldots).$$

PROOF: Because of the fundamental equation (73) (for $n > 0$),

$$\int_0^1 f_n(z)\, dz = \sum_{k=1}^{\infty} \int_0^1 f_{n-1}\left(\frac{1}{k+z} \right) \frac{dz}{(k+z)^2}$$

$$= \sum_{k=1}^{\infty} \int_{\frac{1}{k+1}}^{\frac{1}{k}} f_{n-1}(u)\, du = \int_0^1 f_{n-1}(u)\, du,$$

so that the lemma follows by induction.

PROOF OF THEOREM 33. The function $f_0(x)$ is, by hypothesis, differentiable and hence continuous for $0 \leq x \leq 1$; since it is, again by hypothesis, positive in that interval, it must have some positive minimum, which we shall denote by m. From the condition $m \leq f_0(x) < M$ (for $0 \leq x \leq 1$), we obtain

$$\frac{m}{2(1+x)} < f_0(x) < \frac{2M}{1+x} \qquad (0 \leqslant x \leqslant 1),$$

or

$$\frac{g}{1+x} < f_0(x) < \frac{G}{1+x} \qquad (0 \leqslant x \leqslant 1),$$

where

$$g = \frac{m}{2}, \qquad G = 2M.$$

We now define

$$f_n(x) - \frac{g}{1+x} = \varphi_n(x) \qquad (0 \leqslant x \leqslant 1, \quad n = 0, 1, 2, \ldots).$$

On the basis of Lemma 3, the function $F(x) = g/(1+x)$ satisfies the equation

$$F(x) = \sum_{k=1}^{\infty} F\left(\frac{1}{k+x}\right) \frac{1}{(k+x)^2}$$

(which can easily be shown by direct verification). From this, it obviously follows that the sequence of functions

$$\varphi_0(x), \ \varphi_1(x), \ \ldots, \ \varphi_n(x), \ \ldots$$

satisfies (73). Therefore, all the conclusions that we have deduced from that equation, in particular (74), are valid. As before, we set

$$u = \frac{p_n + x p_{n-1}}{q_n + x q_{n-1}}$$

for brevity. Thus, we have

$$\varphi_n(x) = \sum^{(n)} \varphi_0(u) \frac{1}{(q + x q_{n-1})^2},$$

so that, on the basis of the obvious inequalities

$$q_n + x q_{n-1} \leqslant q_n + q_{n-1} < 2 q_n \qquad \text{and} \qquad \varphi_0(u) > 0,$$

we obtain

$$\varphi_n(x) > \frac{1}{2} \sum^{(n)} \varphi_0(u) \frac{1}{q_n(q_n + q_{n-1})}. \tag{75}$$

On the other hand, the mean value theorem gives us

$$\frac{1}{2} \int_0^1 \varphi_0(z)\,dz = \frac{1}{2} \sum^{(n)} \varphi_0(u') \frac{1}{q_n(q_n + q_{n-1})}, \tag{76}$$

where u' is one of the points of the interval $(p_n/q_n, (p_n + p_{n-1})/(q_n + q_{n-1}))$, and $1/[q_n(q_n + q_{n-1})]$ is the length of this interval. Relations (75) and (76) give us

$$\varphi_n(x) - \frac{1}{2} \int_0^1 \varphi_0(z)\,dz > \frac{1}{2} \sum^{(n)} \{\varphi_0(u) - \varphi_0(u')\} \frac{1}{q_n(q_n + q_{n-1})}.$$

But since, obviously,

$$|\varphi_0(x)| \leqslant |f_0'(x)| + g < \mu + g \qquad (0 \leqslant x \leqslant 1), \tag{77}$$

it follows that

$$|\varphi_0(u) - \varphi_0(u')| < (\mu + g)|u - u'| < \frac{\mu + g}{q_n(q_n + q_{n-1})}$$

$$< \frac{\mu + g}{q_n^2} < \frac{\mu + g}{2^{n-1}}.$$

Inequality (77) then gives us

$$\varphi_n(x) > \frac{1}{2} \int_0^1 \varphi_0(z)\,dz - \frac{\mu + g}{2^n} = l - \frac{\mu + g}{2^n},$$

where

$$l = \frac{1}{2} \int_0^1 \varphi_0(z)\,dz.$$

Thus, we obtain

$$f_n(x) > \frac{g}{1+x} + l - \frac{\mu + g}{2^n} > \frac{g + l - 2^{-n+1}(\mu + g)}{1+x} = \frac{g_1}{1+x}.$$

If we examine the sequence of functions

$$\psi_n(x) = \frac{G}{1+x} - f_n(x) \qquad (n = 0, 1, 2, \ldots)$$

and follow the same line of reasoning as before, we obtain the inequality

$$f_n(x) < \frac{G - l' + 2^{-n+1}(\mu + G)}{1+x} = \frac{G_1}{1+x},$$

where

$$l' = \frac{1}{2} \int_0^1 \psi_0(z)\,dz.$$

Since $l > 0$ and $l' > 0$, we have, for sufficiently large n,

$$g < g_1 < G_1 < G$$

and

$$G_1 - g_1 < G - g - (l + l') + 2^{-n+2}(\mu + G).$$

Also, since

$$l + l' = \frac{1}{2} \int_0^1 \frac{G-g}{1+z}\,dz = (G - g)\frac{\ln 2}{2},$$

we have

$$G_1 - g_1 < (G - g)\delta + 2^{-n+2}(\mu + G),$$

where

$$\delta = 1 - \frac{\ln 2}{2} < 1$$

is an absolute positive constant.

Let us summarize the results that we have obtained. *From the conditions that*

$$\frac{g}{1+x} < f_0(x) < \frac{G}{1+x}, \qquad |f_0'(x)| < \mu,$$

we have shown that, for sufficiently large n,

$$\frac{g_1}{1+x} < f_n(x) < \frac{G_1}{1+x},$$

where

$$g < g_1 < G_1 < G, \quad G_1 - g_1 < (G - g)\delta + 2^{-n+2}(\mu + G).$$

If we now start with the function $f_n(x)$ instead of $f_0(x)$ and repeat the line of reasoning that we have used, we obtain

$$\frac{g_2}{1+x} < f_{2n}(x) < \frac{G_2}{1+x}$$

and

$$g_1 < g_2 < G_2 < G_1,$$
$$G_2 - g_2 < \delta\,(G_1 - g_1) + 2^{-n+2}\,(\mu_1 + G_1),$$

where μ_1 is a positive number such that

$$|f_n'(x)| < \mu_1 \qquad (0 \leqslant x \leqslant 1).$$

If we carry this process further, we obtain, in general,

$$\frac{g_r}{1+x} < f_{rn}(x) < \frac{G_r}{1+x} \qquad (0 \leqslant x \leqslant 1;\ r = 0,\ 1,\ 2,\ \ldots),$$

and, for $r > 0$,

$$\begin{aligned}
g_{r-1} < g_r < G_r < G_{r-1}, \\
G_r - g_r < \delta\,(G_{r-1} - g_{r-1}) + 2^{-n+2}\,(\mu_{r-1} + G_{r-1}),
\end{aligned} \qquad (78)$$

where μ_{r-1} is a positive number such that

$$|f_{(r-1)n}'(x)| < \mu_{r-1} \qquad (0 \leqslant x \leqslant 1).$$

On the basis of Lemma 2, we can set

$$\mu_r = \frac{\mu}{2^{rn-3}} + 4M \qquad (r = 0,\ 1,\ 2,\ \ldots),$$

and, consequently, if n is chosen sufficiently large,

$$\mu_r < 5M \qquad (r = 1,\ 2,\ \ldots).$$

Therefore, successive application of inequality (78) gives us (for $r = 1$, 2, $\cdots$, n)

$$\begin{aligned}
G_n - g_n < (G - g)\,\delta^n + 2^{-n+2}\,\{(\mu + 2M)\,\delta^{n-1} \\
+ 7M\delta^{n-2} + 7M\delta^{n-3} + \ldots + 7M\delta + 7M\}.
\end{aligned}$$

Since $\delta < 1$ is an absolute constant, it obviously follows that

$$G_n - g_n < Be^{-\lambda n},$$

where $\lambda > 0$ is an absolute constant and $B > 0$ depends only on M and μ.

From this, it follows, first of all, that there is a common limit

$$\lim_{n \to \infty} G_n = \lim_{n \to \infty} g_n = a$$

and that

$$\left| f_{n^2}(x) - \frac{a}{1+x} \right| < Be^{-\lambda n} \qquad (0 \leqslant x \leqslant 1), \qquad (79)$$

so that, in particular,

$$\int_0^1 f_{n^2}(z)\,dz \to a \ln 2 \qquad (n \to \infty),$$

and, consequently, on the basis of Lemma 4,

$$a = \frac{1}{\ln 2} \int_0^1 f_0(z)\,dz.$$

Finally, suppose that
$$n^2 \leqslant N < (n+1)^2.$$

Since, on the basis of inequality (79),

$$\frac{a - 2Be^{-\lambda n}}{1+x} < f_{n^2}(x) < \frac{a + 2Be^{-\lambda n}}{1+x},$$

it follows from Lemma 3 that

$$\frac{a - 2Be^{-\lambda n}}{1+x} < f_N(x) < \frac{a + 2Be^{-\lambda n}}{1+x},$$

or

$$\left| f_N(x) - \frac{a}{1+x} \right| < 2Be^{-\lambda n} = Ae^{-\lambda(n+1)} < Ae^{-\lambda \sqrt{N}},$$

where $A = 2Be^{\lambda}$. This inequality, which we have established for sufficiently large N, is obviously true for all $N \geq 0$ if we choose a sufficiently large constant A. This completes the proof of Theorem 33.

Let us now turn to Gauss's problem. If we set

$$f_n(x) = m_n'(x) \qquad (0 \leqslant x \leqslant 1),$$

we obtain $f_0(x) \equiv 1$. Therefore, all the conditions of Theorem 33 are satisfied. Thus, we obtain

$$\left| m_n'(x) - \frac{1}{(1+x)\ln 2} \right| < Ae^{-\lambda \sqrt{n}} \qquad (0 \leqslant x \leqslant 1), \qquad (80)$$

from which, by integrating, we obtain

$$\left| m_n(x) - \frac{\ln(1+x)}{\ln 2} \right| < Ae^{-\lambda \sqrt{n}} \qquad (0 \leqslant x \leqslant 1),$$

where A and λ are absolute positive constants. This not only proves Gauss's assertion, but also gives a good approximation of the remainder term.[6]

Let us now use this result to obtain an approximation of the measure of the set of points for which $a_n = k$ for sufficiently large values of n. Since, obviously, the condition $a_n = k$ is equivalent to the inequalities

$$\frac{1}{k+1} < z_{n-1} \leqslant \frac{1}{k},$$

it follows that

$$\mathfrak{M}E\left(\begin{array}{c} n \\ k \end{array}\right) = m_{n-1}\left(\frac{1}{k}\right) - m_{n-1}\left(\frac{1}{k+1}\right) = \int_{\frac{1}{k+1}}^{\frac{1}{k}} m_{n-1}'(x)\,dx.$$

On the basis of inequality (80), it follows that

$$\left| \mathfrak{M}E\left(\begin{array}{c} n \\ k \end{array}\right) - \frac{\ln\left\{1 + \frac{1}{k(k+2)}\right\}}{\ln 2} \right| < \frac{A}{k(k+1)} e^{-\lambda \sqrt{n-1}}. \qquad (81)$$

We now have a precise limiting relationship for the quantity $\mathfrak{M}E\left(\begin{array}{c} n \\ k \end{array}\right)$,

[6] The method that Lévy used makes it possible to obtain a better approximation. He showed that

$$\left| m_n(x) - \frac{\ln(1+x)}{\ln 2} \right| < Ae^{-\lambda n} \qquad (0 \leqslant x \leqslant 1).$$

(B. G.)

for which we had only rather crude inequalities in section 13. Specifically,

$$\mathfrak{M}E\left(\begin{array}{c} n \\ k \end{array}\right) \rightarrow \frac{\ln\left\{1 + \dfrac{1}{k(k+2)}\right\}}{\ln 2} \qquad (n \rightarrow \infty).$$

Thus, for example, the measure of the set of points for which $a_n = 1$ approaches the quantity

$$\frac{\ln 4 - \ln 3}{\ln 2}$$

as $n \rightarrow \infty$.

Besides proving Gauss's assertion, Theorem 33 enables us to obtain a more general result, the importance of which will be shown below. We denote by $M_n(x)$ the measure of the set of numbers belonging to some fixed interval of rank k and satisfying the condition $z_{k+n} < x$; in other words, $M_n(x)$ is the measure of the set of numbers in the interval $(0, 1)$ satisfying the condition

$$a_1 = r_1, \quad a_2 = r_2, \ldots, \qquad a_k = r_k; \quad z_{k+n} < x, \qquad (82)$$

where $r_1, r_2, \cdots, r_k$ are certain fixed natural numbers and where $n \geq 0$ and $x(0 \leq x \leq 1)$ can be varied arbitrarily.

For the conditions (82) to be satisfied, it is obviously necessary and sufficient that

$$a_1 = r_1, \quad a_2 = r_2, \ldots, \quad a_k = r_k; \qquad \frac{1}{r+x} < z_{k+n-1} \leq \frac{1}{r},$$

where r is some natural number. It then follows that

$$M_n(x) = \sum_{r=1}^{\infty} \left\{ M_{n-1}\left(\frac{1}{r}\right) - M_{n-1}\left(\frac{1}{r+x}\right) \right\} \quad (n \geq 1, \ 0 \leq x \leq 1),$$

so that the sequence of functions $M_0'(x), M_1'(x), \cdots, M_n'(x), \cdots$ satisfies equation (73).

An arbitrary number a of the interval $[p_k/q_k, (p_k + p_{k-1})/(q_k + q_{k-1})]$ can be represented in the form

$$a = \frac{p_k r_{k+1} + p_{k-1}}{q_k r_{k+1} + q_{k-1}},$$

or, since $z_k = 1/r_{k+1}$,

$$\alpha = \frac{p_k + z_k p_{k-1}}{q_k + z_k q_{k-1}}.$$

For $z_k < x$, the number α must lie between p_k/q_k and $(p_k + x p_{k-1})/(q_k + x q_{k-1})$. Therefore,

$$M_0(x) = \left| \frac{p_k}{q_k} - \frac{p_k + x p_{k-1}}{q_k + x q_{k-1}} \right| = \frac{x}{q_k(q_k + q_{k-1}x)}. \qquad (83)$$

If we now set

$$M_n(x) = \mathfrak{M}E \begin{pmatrix} 1, & 2, & \dots, & k \\ r_1, & r_2, & \dots, & r_k \end{pmatrix} \chi_n(x) \qquad (n \geqslant 0, \; 0 \leqslant x \leqslant 1),$$

we obtain a new sequence of functions:

$$\chi_0(x), \; \chi_1(x), \; \dots, \; \chi_n(x), \; \dots;$$

here, the functions $\chi'_n(x)$, which differ from the corresponding functions $M'_n(x)$ only by a constant factor, also satisfy (73). Since obviously

$$\mathfrak{M}E \begin{pmatrix} 1, & 2, & \dots, & k \\ r_1, & r_2, & \dots, & r_k \end{pmatrix} = \left| \frac{p_k}{q_k} - \frac{p_k + p_{k-1}}{q_k + q_{k-1}} \right| = \frac{1}{q_k(q_k + q_{k-1})},$$

equation (83) gives us

$$\chi_0(x) = \frac{(q_k + q_{k-1})x}{q_k + q_{k-1}x},$$

so that

$$\chi'_0(x) = \frac{q_k(q_k + q_{k-1})}{(q_k + q_{k-1}x)^2}$$

and

$$\chi''_0(x) = -\frac{2q_k q_{k-1}(q_k + q_{k-1})}{(q_k + q_{k-1}x)^3}.$$

Thus,

$$\frac{1}{2} < \chi'_0(x) < 2, \quad |\chi''_0(x)| < 4 \qquad (0 \leqslant x \leqslant 1).$$

This shows that Theorem 33 can be applied to the sequence of functions $\chi'_n(x)$, where the numbers A and λ will be arbitrary constants (in particular, independent of $r_1, r_2, \cdots, r_k$). We thus obtain

$$\chi'_n(x) = \frac{M'_n(x)}{\mathfrak{M}E\left(\begin{array}{c} 1,\ 2,\ \ldots,\ k \\ r_1,\ r_2,\ \ldots,\ r_k \end{array}\right)} = \frac{1}{(1+x)\ln 2} + \theta A e^{-\lambda \sqrt{n}}, \quad |\theta| < 1.$$

If we integrate this relation between the limits $1/(r+1)$ and $1/r$, where r is an arbitrary natural number, we obtain, for $|\theta'| < 1$,

$$\frac{M_n\left(\frac{1}{r}\right) - M_n\left(\frac{1}{r+1}\right)}{\mathfrak{M}E\left(\begin{array}{c} 1,\ 2,\ \ldots,\ k \\ r_1,\ r_2,\ \ldots,\ r_k \end{array}\right)} = \frac{\ln\left\{1 + \frac{1}{r(r+2)}\right\}}{\ln 2} + \frac{\theta' A}{r(r+1)} e^{-\lambda \sqrt{n}};$$

and since, obviously,

$$M_n\left(\frac{1}{r}\right) - M_n\left(\frac{1}{r+1}\right) = \mathfrak{M}E\left(\begin{array}{c} 1,\ 2,\ \ldots,\ k,\ k+n+1 \\ r_1,\ r_2,\ \ldots,\ r_k,\ r \end{array}\right),$$

we have

$$\mathfrak{M}E\left(\begin{array}{c} 1,\ 2,\ \ldots,\ k,\ k+n+1 \\ r_1,\ r_2,\ \ldots,\ r_k,\ r \end{array}\right)$$

$$= \left(\frac{\ln\left\{1 + \frac{1}{r(r+2)}\right\}}{\ln 2} + \frac{\theta' A e^{-\lambda \sqrt{n}}}{r(r+1)}\right) \mathfrak{M}E\left(\begin{array}{c} 1,\ 2,\ \ldots,\ k \\ r_1,\ r_2,\ \ldots,\ r_k \end{array}\right).$$

Finally, we can sum this relationship from 1 to ∞ for certain of the numbers $r_1, r_2, \cdots, r_k$ (arbitrarily chosen). As a result of this summation, the terms with like subscripts disappear on both sides of the equation, so that instead of a succession of subscripts 1, 2, $\cdots$, k, we obtain a succession of completely arbitrary subscripts $n_1, n_2, \cdots, n_t$. In all other ways, the equation remains unchanged. Thus, we obtain the following theorem.

THEOREM 34. *Two absolute positive constants A and λ exist such that,* *for* $n_1 < n_2 < \cdots < n_t < n_{t+1}$ *and for arbitrary positive integers* $r_1, r_2, \cdots, r_t, r$,

$$\left| \frac{\mathfrak{M}E \left(\begin{matrix} n_1, & n_2, & \ldots, & n_t, & n_{t+1} \\ r_1, & r_2, & \ldots, & r_t, & r \end{matrix} \right)}{\mathfrak{M}E \left(\begin{matrix} n_1, & n_2, & \ldots, & n_t \\ r_1, & r_2, & \ldots, & r_t \end{matrix} \right)} - \frac{\ln\left\{ 1 + \dfrac{1}{r(r+2)} \right\}}{\ln 2} \right|$$

$$< \frac{A}{r(r+1)} e^{-\lambda \sqrt{n_{t+1} - n_t - 1}}.$$

This result shows not only that the measure of the set of numbers in the interval $(0, 1)$ for which $a_n = r$ approaches a definite limit as $n \to \infty$, but also that the relative measure of the set of numbers satisfying this condition, given arbitrary fixed values of an arbitrary set of preceding elements, approaches the same limit.

16. *Average values*[7]

The results of the preceding section enable us to prove the following general proposition.

THEOREM 35. *Suppose that* $f(r)$ *is a non-negative function of a natural argument* r *and suppose that there exist positive constants* C *and* δ *such that*

$$f(r) < C r^{\frac{1}{2} - \delta} \qquad (r = 1, 2, \ldots).$$

Then, for all numbers in the interval $(0, 1)$, *with the exception of a set of measure zero*,

$$\lim_{n \to \infty} \frac{1}{n} \sum_{k=1}^{n} f(a_k) = \sum_{r=1}^{\infty} f(r) \frac{\ln\left\{ 1 + \dfrac{1}{r(r+2)} \right\}}{\ln 2}. \qquad (84)$$

PRELIMINARY REMARK. The convergence of the series on the right side of (84) follows, of course, from the condition imposed on the function $f(r)$.

PROOF. Let us define

$$\int_0^1 f(a_k) \, d\alpha = u_k, \qquad \int_0^1 \{f(a_k) - u_k\}^2 \, d\alpha = b_k,$$

---

[7] The results of this section are to be found in Khinchin's article "Metrische Kettenbruchprobleme," *Compositio Mathematica*, **1**, 361–382 (1935). (B. G.)

$$\int_0^1 \{f(a_i) - u_i\} \{f(a_k) - u_k\} \, d\alpha = g_{ik},$$

$$\sum_{k=1}^n \{f(a_k) - u_k\} = s_n = s_n(\alpha).$$

The existence of all these integrals follows easily from the properties assumed for the function $f(r)$. For since

$$\{f(r)\}^2 < C^2 r^{1-2\delta},$$

it follows that

$$\int_0^1 \{f(a_k)\}^2 \, d\alpha = \sum_{r=1}^{\infty} \{f(r)\}^2 \, \mathfrak{M}E\binom{k}{r} < 2C^2 \sum_{r=1}^{\infty} \frac{r^{1-2\delta}}{r^2} = C_1$$

is meaningful. The existence of all the integrals then follows on the basis of the Bunyakovskiĭ-Schwarz inequality. In particular, it follows that

$$
\left.
\begin{aligned}
b_k &= \int_0^1 \{f(a_k)\}^2 \, d\alpha - u_k^2 < C_1, \\[2mm]
u_k &= \int_0^1 f(a_k) \, d\alpha < \sqrt{\int_0^1 \{f(a_k)\}^2 \, d\alpha} < \sqrt{C_1}.
\end{aligned}
\right\}
\tag{85}
$$

Furthermore, for $k > i$, we obviously have

$$g_{ik} = \int_0^1 f(a_i) f(a_k) \, d\alpha - u_i u_k$$

$$= \sum_{r,\,s=1}^{\infty} f(r) f(s) \, \mathfrak{M}E\binom{i,\ k}{r,\ s} - u_i u_k. \tag{86}$$

But, on the basis of Theorem 34 and the inequalities of section 12,

$$\left| \mathfrak{M}E\binom{i,\ k}{r,\ s} - \frac{\ln\left\{1 + \dfrac{1}{s(s+2)}\right\}}{\ln 2} \, \mathfrak{M}E\binom{i}{r} \right|$$

$$< \frac{Ae^{-\lambda\sqrt{k-i-1}}}{s(s+1)} \, \mathfrak{M}E\binom{i}{r} \tag{87}$$

$$< 3Ae^{-\lambda\sqrt{k-i-1}} \mathfrak{M}E\binom{i}{r} \mathfrak{M}E\binom{k}{s}$$

and

$$\left| \mathfrak{M}E\begin{pmatrix} k \\ s \end{pmatrix} - \frac{\ln\left\{1 + \dfrac{1}{s(s+2)}\right\}}{\ln 2} \right|$$
$$< \frac{Ae^{-\lambda\sqrt{k-1}}}{s(s+1)} < 3Ae^{-\lambda\sqrt{k-1}}\mathfrak{M}E\begin{pmatrix} k \\ s \end{pmatrix}. \qquad (88)$$

If we multiply inequality (88) by $\mathfrak{M}E\begin{pmatrix} i \\ r \end{pmatrix}$ and compare the result with inequality (87), we obtain

$$\left| \mathfrak{M}E\begin{pmatrix} i, & k \\ r, & s \end{pmatrix} - \mathfrak{M}E\begin{pmatrix} i \\ r \end{pmatrix}\mathfrak{M}E\begin{pmatrix} k \\ s \end{pmatrix} \right|$$
$$< 6Ae^{-\lambda\sqrt{k-i-1}}\mathfrak{M}E\begin{pmatrix} i \\ r \end{pmatrix}\mathfrak{M}E\begin{pmatrix} k \\ s \end{pmatrix},$$

as a result of which (86) gives us

$$\left| g_{ik} - \sum_{r,\,s=1}^{\infty} f(r)f(s)\,\mathfrak{M}E\begin{pmatrix} i \\ r \end{pmatrix}\mathfrak{M}E\begin{pmatrix} k \\ s \end{pmatrix} + u_i u_k \right|$$
$$< 6Ae^{-\lambda\sqrt{k-i-1}} \sum_{r,\,s=1}^{\infty} f(r)f(s)\,\mathfrak{M}E\begin{pmatrix} i \\ r \end{pmatrix}\mathfrak{M}E\begin{pmatrix} k \\ s \end{pmatrix}.$$

Noting that

$$\sum_{r,\,s=1}^{\infty} f(r)f(s)\,\mathfrak{M}E\begin{pmatrix} i \\ r \end{pmatrix}\mathfrak{M}E\begin{pmatrix} k \\ s \end{pmatrix} = u_i u_k$$

and using the second inequality of (85) to estimate the right side of this equation, we obtain

$$|g_{ik}| < 6Ae^{-\lambda\sqrt{k-i-1}}u_i u_k < 6AC_1 e^{-\lambda\sqrt{k-i-1}}. \qquad (89)$$

From (85) and (89), we have, for $n > m > 0$,

$$\int_0^1 (s_n - s_m)^2\, d\alpha = \int_0^1 \left\{ \sum_{k=m+1}^{n} (f(a_k) - u_k) \right\}^2 d\alpha$$
$$= \sum_{k=m+1}^{n} \int_0^1 \{f(a_k) - u_k\}^2\, d\alpha$$

$$+2 \sum_{i=m+1}^{n} \sum_{k=i+1}^{n} \int_0^1 \{f(a_i) - u_i\} \{f(a_k) - u_k\} \, d\alpha \qquad (90)$$

$$= \sum_{k=m+1}^{n} b_k + 2 \sum_{i=m+1}^{n} \sum_{k=i+1}^{n} g_{ik} \cdot < C_1 (n-m)$$

$$+ 12 A C_1 \sum_{i=m+1}^{n} \sum_{k=i+1}^{\infty} e^{-\lambda \sqrt{k-i-1}} < C_2 (n-m),$$

where C_2 is some new positive constant.

We now denote by e_n the set of numbers in the interval $(0, 1)$ for which

$$|s_n| \geqslant \varepsilon n,$$

where ε is an arbitrary given positive constant. Obviously,

$$\int_0^1 s_n^2 \, d\alpha \geqslant \int_{e_n} s_n^2 \, d\alpha \geqslant \varepsilon^2 n^2 \mathfrak{M} e_n,$$

so that inequality (90) (for $m = 0$) gives

$$\mathfrak{M} e_n \leqslant \frac{\int_0^1 s_n^2 \, d\alpha}{\varepsilon^2 n^2} < \frac{C_2}{\varepsilon^2 n}.$$

Thus, the series

$$\sum_{n=1}^{\infty} \mathfrak{M} e_{n^2}$$

converges and, consequently, as we know, almost every number in the interval $(0, 1)$ belongs to only a finite number of sets e_{n^2}, for $n = 1, 2, 3, \cdots$. This means that, for almost all numbers in the interval $(0, 1)$ and for sufficiently large n,

$$\frac{s_{n^2}}{n^2} < \varepsilon;$$

and since ϵ is arbitrarily small, it follows that almost everywhere

$$\lim_{n \to \infty} \frac{s_{n^2}}{n^2} = 0. \tag{91}$$

Furthermore, for $n^2 \leq N < (n+1)^2$, formula (90) gives

$$\int_0^1 (s_N - s_{n^2})\, d\alpha < C_2(N - n^2) < C_2(2n+1) \leqslant 3C_2 n.$$

If we denote by $e_{n,N}$ the set of numbers in the interval $(0, 1)$ for which $|s_N - s_{n^2}| \geq \epsilon n^2$ and if we set

$$\sum_{N=n^2}^{(n+1)^2-1} e_{n, N} = E_n,$$

we then have, for $n^2 \leq N < (n+1)^2$,

$$\int_0^1 (s_N - s_{n^2})^2\, d\alpha \gg \int_{e_{n, N}} (s_N - s_{n^2})^2\, d\alpha > \epsilon^2 n^4 \mathfrak{M} e_{n, N},$$

$$\mathfrak{M} e_{n, N} < \frac{3C_2}{\epsilon^2 n^3},$$

$$\mathfrak{M} E_n \leqslant \sum_{N=n^2}^{(n+1)^2-1} \mathfrak{M} e_{n, N} < \frac{3C_2(2n+1)}{\epsilon^2 n^3} \leqslant \frac{9C_2}{\epsilon^2 n^2},$$

so that the series $\sum_{n=1}^{\infty} \mathfrak{M} E_n$ converges. Almost every number in the interval $(0, 1)$ must then belong to only a finite number of sets E_n and, hence, to only a finite number of sets $e_{n,N}$. But this means that almost all numbers in the interval $(0, 1)$ satisfy the inequality

$$|s_N - s_{n^2}| < \epsilon n^2,$$

for sufficiently large n and for $n^2 \leq N < (n+1)^2$. In other words, almost everywhere

$$\frac{|s_N - s_{n^2}|}{n^2} < \epsilon,$$

for sufficiently large n and for $n^2 \leq N < (n+1)^2$. Since ϵ is arbitrarily small, it follows that almost everywhere

$$\frac{s_N}{n^2} - \frac{s_{n^2}}{n^2} \to 0 \quad [n \to \infty, \ n^2 \leqslant N < (n+1)^2].$$

On the basis of equation (91), it then follows that almost everywhere

$$\frac{s_N}{n^2} \to 0 \quad [n \to \infty, \ n^2 \leqslant N < (n+1)^2],$$

and hence, a fortiori,

$$\frac{s_N}{N} \to 0 \quad (N \to \infty).$$

In other words, almost everywhere,

$$\frac{1}{N} \sum_{k=1}^{N} f(a_k) - \frac{1}{N} \sum_{k=1}^{N} u_k \to 0 \quad (N \to \infty). \tag{92}$$

But from formula (81) of the preceding section

$$\left| u_k - \sum_{r=1}^{\infty} f(r) \frac{\ln \left\{ 1 + \dfrac{1}{r(r+2)} \right\}}{\ln 2} \right| = \sum_{r=1}^{\infty} f(r) \left| \mathfrak{M}E\left(\frac{k}{r}\right) \right|$$

$$\frac{\ln\left(1 + \dfrac{1}{r(r+2)}\right)}{\ln 2} \bigg| < A e^{-\lambda \sqrt{k-1}} \sum_{r=1}^{\infty} \frac{f(r)}{r(r+1)} < A_1 e^{-\lambda \sqrt{k}},$$

where A_1 is a new positive constant. Therefore,

$$u_k \to \sum_{r=1}^{\infty} f(r) \frac{\ln\left\{1 + \dfrac{1}{r(r+2)}\right\}}{\ln 2} \quad (k \to \infty),$$

and, consequently,

$$\frac{1}{N} \sum_{k=1}^{N} u_k \to \sum_{r=1}^{\infty} f(r) \frac{\ln\left\{1 + \dfrac{1}{r(r+2)}\right\}}{\ln 2} \quad (N \to \infty).$$

Relation (92) then gives us

$$\frac{1}{N}\sum_{k=1}^{N} f(a_k) \to \sum_{r=1}^{\infty} f(r)\, \frac{\ln\left\{1 + \dfrac{1}{r(r+2)}\right\}}{\ln 2}$$

almost everywhere in the interval $(0, 1)$. This completes the proof of Theorem 35.

This theorem enables us to establish quite a number of properties of continued fractions that are satisfied for almost all irrational numbers. For example, let us set

$$f(r) = 1, \text{ for } r = k \quad \text{and} \quad f(r) = 0, \text{ for } r \neq k,$$

where k is some (arbitrary) natural number. In this case, the sum

$$\psi_n(k) = \sum_{i=1}^{n} f(a_i)$$

obviously represents the number of times the integer k occurs among the first n elements of a given continued fraction. On the other hand, the ratio

$$\frac{\psi_n(k)}{n} = \frac{1}{n}\sum_{i=1}^{n} f(a_i)$$

gives us the density of the number k among the first n elements of the given continued fraction. Finally, the limit

$$\lim_{n\to\infty} \frac{\psi_n(k)}{n} = d(k),$$

if it exists, is naturally interpreted as the density of the number k in the entire sequence of elements of the given continued fraction.

Since the function $f(r)$ that we have defined clearly satisfies all of the requirements of Theorem 35, we conclude, on the basis of that theorem, that, *for arbitrary k, this density exists almost everywhere and that it has the same value almost everywhere.* Furthermore, the same theorem makes it possible for us to calculate the value of that density. Obviously, we have almost everywhere

$$d(1) = \frac{\ln 4 - \ln 3}{\ln 2}, \quad d(2) = \frac{\ln 9 - \ln 8}{\ln 2}, \quad d(3) = \frac{\ln 16 - \ln 15}{\ln 2},$$

and so on. Thus, an arbitrary natural number occurs as an element in the expansion of almost all numbers with equal average frequency.

We obtain another interesting result by setting

$$f(r) = \ln r \qquad (r = 1,\ 2,\ 3,\ \ldots).$$

All the conditions of Theorem 35 are then satisfied. Therefore, we see that almost everywhere

$$\frac{1}{n}\sum_{i=1}^{n}\ln a_i \to \sum_{r=1}^{\infty}\ln(r)\,\frac{\ln\left\{1+\dfrac{1}{r(r+2)}\right\}}{\ln 2} \qquad (n \to \infty),$$

or, equivalently,

$$\sqrt[n]{a_1 a_2\ \ldots\ a_n} \to \prod_{r=1}^{\infty}\left\{1+\frac{1}{r(r+2)}\right\}^{\frac{\ln r}{\ln 2}}$$

Thus, the geometric mean of the first n elements approaches the absolute constant

$$\prod_{r=1}^{\infty}\left\{1+\frac{1}{r(r+2)}\right\}^{\frac{\ln r}{\ln 2}} = 2,6\ \ldots,$$

almost everywhere as $n \to \infty$.

Obviously, Theorem 35 makes it possible to establish analogous results for a whole series of other types of average values. However, investigation of the arithmetic mean

$$\frac{1}{n}\sum_{i=1}^{n}a_i \tag{93}$$

by this method is impossible, because the corresponding function $f(r) = r$ does not satisfy the conditions of Theorem 35. However, it is easy to see from more elementary considerations that, almost everywhere, the expression (93) cannot have any kind of finite limit. For Theorem 30 (sec. 13) tells us that almost everywhere

$$a_n > n \ln n$$

for an infinite number of values of n, and hence, a fortiori,

$$\sum_{i=1}^{n} a_i > n \ln n, \text{ and hence, } \frac{1}{n} \sum_{i=1}^{n} a_i > \ln n.$$

Thus, the quantity (93) is almost everywhere unbounded and therefore, as we stated, cannot have a finite limit.

INDEX

A CATALOG OF SELECTED

DOVER BOOKS
IN SCIENCE AND MATHEMATICS

Mathematics

FUNCTIONAL ANALYSIS (Second Corrected Edition), George Bachman and Lawrence Narici. Excellent treatment of subject geared toward students with background in linear algebra, advanced calculus, physics and engineering. Text covers introduction to inner-product spaces, normed, metric spaces, and topological spaces; complete orthonormal sets, the Hahn-Banach Theorem and its consequences, and many other related subjects. 1966 ed. 544pp. 6⅛ x 9¼. 0-486-40251-7

DIFFERENTIAL MANIFOLDS, Antoni A. Kosinski. Introductory text for advanced undergraduates and graduate students presents systematic study of the topological structure of smooth manifolds, starting with elements of theory and concluding with method of surgery. 1993 edition. 288pp. 5⅜ x 8½. 0-486-46244-7

VECTOR AND TENSOR ANALYSIS WITH APPLICATIONS, A. I. Borisenko and I. E. Tarapov. Concise introduction. Worked-out problems, solutions, exercises. 257pp. 5⅝ x 8¼. 0-486-63833-2

AN INTRODUCTION TO ORDINARY DIFFERENTIAL EQUATIONS, Earl A. Coddington. A thorough and systematic first course in elementary differential equations for undergraduates in mathematics and science, with many exercises and problems (with answers). Index. 304pp. 5⅜ x 8½. 0-486-65942-9

FOURIER SERIES AND ORTHOGONAL FUNCTIONS, Harry F. Davis. An incisive text combining theory and practical example to introduce Fourier series, orthogonal functions and applications of the Fourier method to boundary-value problems. 570 exercises. Answers and notes. 416pp. 5⅜ x 8½. 0-486-65973-9

COMPUTABILITY AND UNSOLVABILITY, Martin Davis. Classic graduate-level introduction to theory of computability, usually referred to as theory of recurrent functions. New preface and appendix. 288pp. 5⅜ x 8½. 0-486-61471-9

AN INTRODUCTION TO MATHEMATICAL ANALYSIS, Robert A. Rankin. Dealing chiefly with functions of a single real variable, this text by a distinguished educator introduces limits, continuity, differentiability, integration, convergence of infinite series, double series, and infinite products. 1963 edition. 624pp. 5⅜ x 8½. 0-486-46251-X

METHODS OF NUMERICAL INTEGRATION (SECOND EDITION), Philip J. Davis and Philip Rabinowitz. Requiring only a background in calculus, this text covers approximate integration over finite and infinite intervals, error analysis, approximate integration in two or more dimensions, and automatic integration. 1984 edition. 624pp. 5⅜ x 8½. 0-486-45339-1

INTRODUCTION TO LINEAR ALGEBRA AND DIFFERENTIAL EQUATIONS, John W. Dettman. Excellent text covers complex numbers, determinants, orthonormal bases, Laplace transforms, much more. Exercises with solutions. Undergraduate level. 416pp. 5⅜ x 8½. 0-486-65191-6

RIEMANN'S ZETA FUNCTION, H. M. Edwards. Superb, high-level study of landmark 1859 publication entitled "On the Number of Primes Less Than a Given Magnitude" traces developments in mathematical theory that it inspired. xiv+315pp. 5⅜ x 8½. 0-486-41740-9

CATALOG OF DOVER BOOKS

A TREATISE ON ELECTRICITY AND MAGNETISM, James Clerk Maxwell. Important foundation work of modern physics. Brings to final form Maxwell's theory of electromagnetism and rigorously derives his general equations of field theory. 1,084pp. 5⅜ x 8½. Two-vol. set. Vol. I: 0-486-60636-8 Vol. II: 0-486-60637-6

MATHEMATICS FOR PHYSICISTS, Philippe Dennery and Andre Krzywicki. Superb text provides math needed to understand today's more advanced topics in physics and engineering. Theory of functions of a complex variable, linear vector spaces, much more. Problems. 1967 edition. 400pp. 6½ x 9¼. 0-486-69193-4

INTRODUCTION TO QUANTUM MECHANICS WITH APPLICATIONS TO CHEMISTRY, Linus Pauling & E. Bright Wilson, Jr. Classic undergraduate text by Nobel Prize winner applies quantum mechanics to chemical and physical problems. Numerous tables and figures enhance the text. Chapter bibliographies. Appendices. Index. 468pp. 5⅜ x 8½. 0-486-64871-0

METHODS OF THERMODYNAMICS, Howard Reiss. Outstanding text focuses on physical technique of thermodynamics, typical problem areas of understanding, and significance and use of thermodynamic potential. 1965 edition. 238pp. 5⅜ x 8½. 0-486-69445-3

THE ELECTROMAGNETIC FIELD, Albert Shadowitz. Comprehensive undergraduate text covers basics of electric and magnetic fields, builds up to electromagnetic theory. Also related topics, including relativity. Over 900 problems. 768pp. 5⅝ x 8¼. 0-486-65660-8

GREAT EXPERIMENTS IN PHYSICS: FIRSTHAND ACCOUNTS FROM GALILEO TO EINSTEIN, Morris H. Shamos (ed.). 25 crucial discoveries: Newton's laws of motion, Chadwick's study of the neutron, Hertz on electromagnetic waves, more. Original accounts clearly annotated. 370pp. 5⅜ x 8½. 0-486-25346-5

EINSTEIN'S LEGACY, Julian Schwinger. A Nobel Laureate relates fascinating story of Einstein and development of relativity theory in well-illustrated, nontechnical volume. Subjects include meaning of time, paradoxes of space travel, gravity and its effect on light, non-Euclidean geometry and curving of space-time, impact of radio astronomy and space-age discoveries, and more. 189 b/w illustrations. xiv+250pp. 8⅜ x 9¼. 0-486-41974-6

THE VARIATIONAL PRINCIPLES OF MECHANICS, Cornelius Lanczos. Philosophic, less formalistic approach to analytical mechanics offers model of clear, scholarly exposition at graduate level with coverage of basics, calculus of variations, principle of virtual work, equations of motion, more. 418pp. 5⅜ x 8½. 0-486-65067-7